MRS. S.C. HALL –
A LITERARY BIOGRAPHY

THE IRISH LITERARY STUDIES SERIES
ISSN 0140-895X

1. *Place, Personality & the Irish Writer.* Andrew Carpenter (editor)
2. *Yeats and Magic.* Mary Catherine Flannery
3. *A Study of the Novels of George Moore.* Richard Allen Cave
4. *J. M. Synge & the Western Mind.* Weldon Thornton
5. *Irish Poetry from Moore to Yeats.* Robert Welch
6. *Yeats, Sligo & Ireland.* A. Norman Jeffares (editor)
7. *Sean O'Casey, Centenary Essays.* David Krause & Robert G. Lowery (editors)
8. *Denis Johnston: A Retrospective.* Joseph Ronsley (editor)
9. *Literature & the Changing Ireland.* Peter Connolly (editor)
10. *James Joyce: An International Perspective.* Suheil Badi Bushrui & Bernard Benstock (editors)
11. *Synge: the Medieval & the Grotesque.* Toni O'Brien Johnson
12. *Carleton's Traits and Stories & the 19th Century Anglo-Irish Tradition.* Barbara Hayley
13. *Lady Gregory: Fifty Years After.* Ann Saddlemyer & Colin Smythe (editors)
14. *Women in Irish Legend, Life & Literature.* Edited by S. F. Gallagher (editor)
15. *'Since O'Casey' & Other Essays on Irish Drama.* Robert Hogan
16. *George Moore in Perspective.* Janet Egleson Dunleavy (editor)
17. *W. B. Yeats, Dramatist of Vision.* A. S. Knowland
18. *The Irish Writer & the City.* Maurice Harmon (editor)
19. *O'Casey the Dramatist.* Heinz Kosok
20. *The Double Perspective of Yeats's Aesthetic.* Okifumi Komesu
21. *The Pioneers of Anglo-Irish Fiction, 1800–1850.* Barry Sloan
22. *Irish Writers & Society at Large.* Masaru Sekine (editor)
23. *Irish Writers & the Theatre.* Masaru Sekine (editor)
24. *A History of Verse Translation from the Irish 1789–1897.* Robert Welch
25. *Kate O'Brien, A Literary Portrait.* Lorna Reynolds
26. *Portraying the Self, Sean O'Casey and the Art of Autobiography.* Michael Kenneally
27. *W. B. Yeats & the Tribes of Danu.* Peter Alderson Smith
28. *Theatre of Shadows: Samuel Beckett's Drama 1956–76.* Rosemary Pountney
29. *Critical Approaches to Anglo-Irish Literature.* Michael Allen & Angela Wilcox (editors)
30. *'Make Sense Who May': Essays on Samuel Beckett's Later Works.* Robin J. Davis & Lance St. J. Butler (editors)
31. *Cultural Contexts and Literary Idioms in Contemporary Irish Literature.* M. Kenneally (editor)
32. *Builders of My Soul: Greek and Roman Themes in Yeats.* Brian Arkins
33. *Perspectives of Irish Drama and Theatre.* Jacqueline Genet & Richard Allen Cave (editors)
34. *The Great Queens. Irish Goddesses from the Morrigan to Cathleen ni Houlihan.* Rosalind Clark
35. *Irish Literature and Culture.* Michael Kenneally (editor)
36. *Irish Writers and Politics.* Okifumi Komesu & Masaru Sekine (editors)
37. *Irish Writers and Religion.* Robert Welch (editor)
38. *Yeats and the Noh.* Masaru Sekine & Christopher Murray
39. *Samuel Ferguson: the Literary Achievement.* Peter Denman
40. *Reviews and Essays of Austin Clarke.* Gregory A. Schirmer (editor)
41. *The Internationalism of Irish Literature & Drama.* Joseph McMinn (editor)
42. *Ireland and France, A Bountiful Friendship: Literature, History and Ideas.* Barbara Hayley & Christopher Murray (editors)
43. *Poetry in Contemporary Irish Literature.* Michael Kenneally (editor)
44. *International Aspects of Irish Literature.* Toshi Furomoto & George Hughes, et al. (editors)
45. *A Small Nation's Contribution to the World.* Donald E. Morse, Csilla Bertha, István Pàllffy (editors)
46. *Images of Invention. Essays on Irish Writing.* A. Norman Jeffares
47. *Literary Inter-Relations: Ireland, Egypt, and the Far East.* Mary Massoud (editor)
48. *Irish Writers and their Creative Process.* Jacqueline Genet & Wynn Hellegouarc'h (editors)
49. *Rural Ireland, Real Ireland?* Jacqueline Genet (editor)
50. *Mrs S. C. Hall, A Literary Biography.* Maureen Keane

MRS. S.C. HALL
A Literary Biography

Maureen Keane

Irish Literary Studies: 50

COLIN SMYTHE
Gerrards Cross, 1997

Copyright © 1997 by Maureen Keane

The right of Maureen Keane to be identified as the Author of this work has been asserted in accordance with the Copyright, Designs and Patents Act, 1998

All rights reserved. Apart from any fair dealing for the purposes of research or private study, or criticism or review, as permitted under the Copyright, Designs and Patents Act, 1988, this publication may be reproduced, stored or transmitted, in any forms or by any means, only with the prior permission in writing of the publishers, or in the case of reprographic reproduction in accordance with the terms of licences issued by the Copyright Licensing Agency. Inquiries concerning reproduction outside these terms should be sent to the publishers at the undermentioned address.

First published in Great Britain in 1997
by Colin Smythe Limited, Gerrards Cross,
Buckinghamshire SL9 8XA

British Library Cataloguing in Publication Data

A catalogue record for this book is available from the British Library

ISBN 0-86140-394-0

Distributed in North America by Oxford University Press
198 Madison Avenue, New York, NY 10016

Produced in Great Britain
Printed and bound by T.J. International Ltd
Padstow, Cornwall

IN MEMORY OF BARBARA HAYLEY

Contents

INTRODUCTION. ix

ACKNOWLEDGEMENTS. xi

IRELAND – 'THE GREAT MART OF FICTION'. 1

MRS HALL – MARRIAGE AND MARKETS. 17

TEACHING – THE TASTE OF THE TIMES. 35

SKETCHES OF IRISH LIFE – THE VOICE OF THE COLONIST. 51

LIGHTS AND SHADOWS – A MELANCHOLY BOOK. 75

STORIES OF THE IRISH PEASANTRY – CORRECTING THE 'EVIL HABITS OF POOR PAT'. 95

HALLS' IRELAND – 'GUIDANCE FOR THOSE WHO DESIGN TO VISIT IRELAND. 113

THE WHITEBOY – 'A TRULY NATIONAL NOVEL'. 145

THREE NOVELISTS WITH A COMMON CAUSE. 177

ASSESSMENTS – THEN AND NOW. 199

BIBLIOGRAPHY. 215

NOTES. 233

INDEX. 251

Introduction

In April 1829 a collection of Irish stories was published in London by Westley and Davis. Called *Sketches of Irish Character*[1], the two little volumes contained 11 tales by a Mrs S.C. Hall, a lady who, although living in London, had been born and brought up in Ireland and who proudly proclaimed herself Irish. The work was an immediate success and ran into a second edition. Two years later a Second Series of *Sketches of Irish Character*[2] appeared and this further collection of stories was even more popular with the public and the critics. The First and Second series of the *Sketches* were later published in combined editions, the last British publication being in 1913. Mrs Hall had been accepted from the beginning as an expert on Irish life and Irish character, and her reputation was enhanced by the publication of *Lights and Shadows of Irish Life*[3] in 1838, a more reflective work which went beyond mere story-telling. In *Stories of the Irish Peasantry*[4] which appeared in 1840, Mrs Hall illustrated for her English and Scottish readers those faults which she found to be exclusively Irish. Mrs Hall's position as guide to the complexities of Irish life was finally established for her readers by the appearance of the work she wrote in collaboration with her husband Samuel Carter Hall, a London journalist, – *Ireland, its Scenery, Character, etc.*[5] This was published first in weekly parts and then in three handsomely-bound volumes by How and Parsons in 1843. A full-length novel, *The Whiteboy: A Story of Ireland in 1822*[6], set in Ireland and dealing with Irish problems, political and social, was published in 1845. Another novel, *Midsummer Eve: A Fairy Tale of Love*[7], set in Ireland, appeared in 1847, but made little impact. *The Whiteboy* was Mrs Hall's last significant work about Ireland and although she later wrote a novel, *The Fight of Faith*[8] in 1869 and occasional Irish tales and pamphlets, they never achieved the popularity of the earlier works.

In addition to her Irish stories Mrs Hall wrote novels with an English setting, historical and contemporary, guide books to English beauty spots and places of pilgrimage, numerous stories for children and an amazing amount of journalism, including editorial work on various magazines and periodicals. Of all that prodigious output the Irish stories alone stand the test of time and are the only ones of any literary merit. They were also, in her own lifetime, the most popular, and the most highly regarded. She was taken at her own valuation as a writer who could explain some of the mysteries of the Irish character and her picture of Ireland was accepted abroad as the truthful one. Even in Ireland she was regarded, in some circles, as a true interpreter of Irish life. What she wrote not only reflected a certain attitude towards the Irish people, it helped to influence English public opinion. It is timely, so, to examine her work, and to explore the reasons for its success.

Acknowledgements

My thanks are due to the librarians and staff of the National Library of Ireland, Trinity College Dublin, the British Library, and the British Newspaper Library at Colindale.

My publisher, Colin Smythe, is to be thanked for his cheerful encouragement and guidance, and I will always be grateful to Helen and Geoffrey Parker for their hospitality in London.

Frances O'Shea, who typed this work, displayed not only impressive professional skills but proved to be a true, patient and understanding friend, ready and willing to cope with any crisis. I thank her most sincerely.

Finally, I thank my husband, David Keane, who has given me every possible support over the years.

ONE

Ireland – 'The Great Mart of Fiction'

Anna Maria Fielding, later Mrs Samuel Carter Hall, who was born in Dublin in 1800, was brought up in a well-to-do, comfortable home in County Wexford. This, as it turned out, was an excellent background for an Irish writer living and working in London. Ireland, although geographically close, could fairly be described as a foreign land and accounts of foreign travels were popular in England in a way they had never been before. English travellers had, of course, always written accounts of their voyages, and the professional explorers such as Captain Cook, Captain Bligh, Captain Vancouver, Mungo Park and James Bruce published tales whose chief value lay in their intrinsic novelty. Then there were accounts which were by-products of a major theme; for example those of John Wesley on his mission and Arthur Young on agriculture. A new species of travel writer, however, began to appear at the beginning of the nineteenth century. This was the ordinary person, man or woman, who travelled the world and wrote an account of the journey in the form of sketches, memoirs, diaries and journals. Some of these people travelled for the sake of travelling and for the pleasure to be gained from it (not 'pleasure' in the sense a stay-at-home would recognise since it often arose from surviving danger and discomfort), and for the express purpose of writing about it. Publishers began to commission travel books, as did Henry Colburn for Lady Morgan's *France* in 1816.

The literary magazines which had made their appearance since 1800 helped to popularise travel writing by their notices and reviews. The reviews were often little more than long extracts from the works in question, prefaced by a few perfunctory words of recommendation. An extremely large number of these works was noted; for instance between October 1805 and October 1807 twenty-four travel books were reviewed in the *Edinburgh Review*, including works as diverse as Rainsford's *Account of Haiti*, Barrow's *Account of a Journey in Africa*, Kotzebue's *Travels in Italy*, Carr's *Stranger in Ireland* and Turnbull's *Voyage Round the World*. From February 1815 to February 1818 there was a decline in the number of works reviewed but the space allotted was still considerable. During this period Lewis and Clark's *Travels to the Source of the Missouri,* Holland's *Travels in Albania,* Stendhal's *Rome, Naples and Florence,* and Captain Basil Hall's *Voyage to Loo-Choo* were among works noted and extensively quoted. Between March 1826 and July 1827 *Blackwood's Edinburgh Magazine* had articles and reviews dealing with travel in Ireland, Central Africa, America, the Spanish Peninsula and India. There were also articles on American ornithology and on the way of life of women in Persia. *Fraser's Magazine for Town and Country* led off Volume 1 in 1830 with West Indian sketches, Canadian sketches, and reviews of Webster's *Travels through India and the Crimea* and Burckhardt's *Arabic Proverbs.* In Volume II, among others, were reviews of Mackenzie's *Haiti* and Bayley's *Four Years in the West Indies,* and Volume III had a long and unfavourable review of Lady Morgan's *France* in 1829-30.

Descriptions of landscape were important features of most of these travel books and the current fascination with wild and rugged scenery is fully reflected in much of the writing. The great debate about what was beautiful and what was sublime continued long after Edmund Burke first made the distinction in his *Philosophic Enquiry into the Origin of Our Ideas of the Sublime and the Beautiful* in 1757 but it was generally accepted that the quality of ruggedness was one of the attributes of the sublime. This ruggedness could also be described as 'picturesque' in its literal meaning of being like a picture. Not just any picture but one which depicted savage and untamed scenery. The most celebrated painter of this type of scene was Salvator Rosa, and his romantic vision brought forth trees that waved tortured branches against wild skies, rocks poised to tumble down precipices, and cataracts thundering down jagged crags.

The opposing type of beauty, that of classical serenity, was exemplified in the paintings of Claude Lorraine and Nicholas Poussin. Claude's paintings are static, the embodiment of the contentment that comes from reason and philosophic meditation, and Poussin too conveys a sense of reason and of logic, although there is an underlying menace and a sense of foreboding in even the most classically composed of his works. These three painters, Nicolas Poussin (1594–1665), Claude Lorraine (1600–1682), and Salvator Rosa (1615–1673) between them illustrate the aesthetic tastes of the age, and it is not surprising that the literary descriptions of landscape should have been influenced by the styles of the most popular painters. In her work *Italian Landscape in Eighteenth Century England,* a *Study Chiefly of the Influence of Claude Lorraine and Salvator Rosa on English Taste,*[1] Elizabeth Manwaring cites a number of examples to prove this point and quotes from M.G. Lewis, Lady Charlotte Bury, G.P.R. James, Mrs Ann Radcliffe, William Godwin, and Sir Walter Scott, passages which have been influenced by Poussin, Lorraine, and Rosa. Of the three painters it is Rosa who seems to have had the greatest influence on English and Scottish writers and it is certainly Rosa who most appealed to Irish writers. Passages in the works of Lady Morgan, the Banim brothers, Gerald Griffin and William Carleton all testify to his fascination for Irish writers (although Carleton has some purely Claudian compositions, which, interestingly enough, have a much fresher, more personal air than his rather hackneyed Rosa-influenced landscape descriptions).

Mrs Hall was a fervent admirer of Salvator Rosa. She called him her 'king of painters', and in her short story 'The Rapparee' in the Second Series of *Sketches of Irish Life and Character*,[2] published in 1831, she states that the setting for the ambush by highwaymen was 'one that Salvator only could have painted'. She goes on to describe it: 'the ground descended steeply, but unevenly, into a hollow glen, one side of which was skirted by stunted and straggling brushwood, that fringed what was called the carriage road, while the other sloped down to a sort of shingly bottom (the black glen), through which a mountain stream brawled angrily and restlessly on its way'.[3] It is indeed a scene straight from a Rosa painting and the highwaymen are the Irish equivalent of the banditti whom Rosa featured in his paintings. In her Irish novel *The Whiteboy: A Story of Ireland in 1822*,[4] although it was written and published much later, the influence of Rosa is still apparent, particularly in a scene where the Irishman, Louis O'Brien, Captain in the French army, and liaison officer with the Irish rebels planning revolt, has his hideout in Glen Flesk in 'the fissure of a rock which commanded an extensive view.' He, too, is an Irish bandit and he sees in front of him a typical Salvatorean tree, blasted and burnt by lightning. 'Adamantine rocks' towered, and torrents were 'rushing down the hills brawling and wrestling on their way; leaping in sport fathom-deep from crag to crag in mimic cataracts'.[5] In *Ireland, its Scenery, Character, etc.*,[6] a guide book which Mrs Hall wrote in collaboration with her husband Samuel Carter Hall, fact and fiction were blended, and from the style of the stories and the prose used to describe scenery it is fair to assume that Mrs Hall was responsible for those elements. The description of Glendalough, for instance, has affinities with the extract from *The Whiteboy* quoted above – 'descending rivulets', combining to make the scene 'awfully grand'.

This magnificent scenery of Ireland, so eloquently described by Mrs Hall in the style and vocabulary favoured at the time was, of course, wonderful for admirers of the sublime and the beautiful to read about, and obviously Ireland was truly picturesque, but that alone would not account for its appeal as a foreign land. After all, such natural wonders could be found nearer home, in England itself, in Scotland and in Wales. The climate was much the same, lacking in extremes of heat and cold, the inhabitants speaking a version of English that could, with a little effort and much laughter, be quite easily understood; they owed allegiance, theoretically at least, to the same Crown, were governed by the same laws, and had a monotheistic religion, perverted though it was by Popery. Earlier travellers had assured readers that journeying in Ireland was safe, that there was a high standard of sexual morality there, and that hospitality to strangers was almost a code of honour, yet there were still doubts about the country. An aura of violence hung about it, and the Rebellion of 1798 had proved within living memory that Ireland was not a truly peaceful place. The *Eclectic Review*, in 1810 reviewing *Tahetian* [sic] *and Irish Spelling Books* stated that the Irish, compared with the rest of Europe, were 'barbarous' and that the 'common Irish have always been so'.[7] That barbarity had manifested itself chiefly in Wexford in 1798 and there were well-attested eye-witness accounts of rebel atrocities and attacks upon the Protestant residents there.

Now came a calming voice out of Ireland. Anna Maria Hall, wife of a respected London journalist, could set fears about Ireland at rest. Her first *Sketches of Irish Character*[8] 1829, was set entirely in Wexford and the Second Series of *Sketches* in 1831 mainly so. Yet this was the area in which the rebellion had been at its fiercest and where so many good Protestants had been slain. Nonetheless, Mrs Hall was able to report on a region where resident landlords lived in peace and amity with their tenants, and where even the most recent disturbances were now, if not forgotten, no longer agitating men's minds with thoughts of sedition. Mrs Hall's own family had had first-hand experience of the turmoil of the rebellion but she was able to report how they lived in safety thereafter. One of her stories in the first series of the *Sketches* concerns an elderly Roman Catholic priest who had been involved in the uprising but who later regretted it (as did another eponymous hero, Andy the Miller) and was glad that he had helped save the life of some Protestant friends. They, in their turn, helped him evade the consequences of his action in joining the rebels. This story 'Father Mike', was based on an experience Mrs Hall's grandmother had had in 1798. In the introduction to the Fifth Edition of the *Sketches*, published in 1854, Mrs Hall told of her 'stately grandmother' who 'for all her

Huguenot feeling' was on very friendly terms with the local Roman Catholic parish priest who 'had saved her and hers in the time of the Rebellion of '98 from destruction'. Then,

> when the time of retribution came, and the priests of the County Wexford were under the ban of government, she wrote a letter to a person "having authority," representing his care of her and her daughter, and his exertions to save the property and life of her husband, and in due time the Bannow priest was told he had nothing to fear; thus they were bound to each other.[9]

Such stories were extremely reassuring, based as they were on personal experience. In the Introduction to the Third Edition of the *Sketches* in 1842 Mrs Hall emphasised the autobiographical nature of the tales and reminded readers that they chiefly referred to 'one locality – the parish of Bannow, in the sea-coast of the County of Wexford' her own native place, where she had spent the earlier years of her life. The county of Wexford, she points out, has many 'moral, social and natural advantages', and her native area, the Baronies of Bargy and Forth, is even more richly endowed with these advantages:

> The inhabitants are, chiefly, descendants of the Anglo-Norman settlers, who, in the reign of the second Henry, invaded and conquered – or, rather, subdued – Ireland; ... . The people are to this day, "a peculiar people," and retain much of their English character. This is apparent, not alone in the external aspect of the country – in the skilfully farmed fields, the comparatively comfortable cottages, the barns attached to every farmyard, the well-trimmed hedgerows, stocked with other vegetables than potatoes; the peasantry are better clad than we have seen them elsewhere, and have an air of sturdy independence, which they really feel, and to which they are justly entitled, for it is achieved by their own honest industry.[10]

It was unnecessary for Mrs Hall to spell out her personal involvement in this way, because almost every story in the *Sketches*, especially in the First Series, shows a very close acquaintanceship with a district and its people. Carefully-noted details of domestic life and dress, and an awareness of rhythms of speech and an insight into certain aspects of Irish peasant character recreate a living world. This was the world in which Mrs Hall grew up.

In his autobiography, *Retrospect of a Long Life, from 1815 to 1883*,[11] Samuel Carter Hall, Mrs Hall's husband, states that his wife was born on 6th January, 1800 in South Anne St., Dublin. Her mother, a Mrs Fielding, was left a widow shortly after the child's birth, and she moved with the little girl to stay with her own mother and stepfather, Mr and Mrs George Carr, a well-to-do Protestant couple, on

George Carr's estate at Graige, in Bannow, County Wexford. It was here that Anna Maria grew up, and it was here that she received her earliest impressions. It was a happy childhood in comfortable circumstances, and the Carrs were popular locally among the Catholic tenantry. Mrs Carr, according to Mr Hall, was a lady of Huguenot descent, whose forbears had fled France after the Revocation of the Edict of Nantes. Widowed young and left, as her daughter, Mrs Fielding later was, with a young child, she married George Carr when he was on a visit to England. Mr Hall records that this lady died in 1815 and that George Carr, Mrs Fielding and Anna Maria then left Bannow and went to live in London. He gives no more information about George Carr, who is buried in Bannow churchyard with date of death given as 1824, but a memoir written by a relative of George Carr's, Richard Boyse Osborne, tells an interesting story. George Carr was Osborne's grand-uncle and inherited the large estate of Graige when he was aged about 39, in 1783. The annual income from the estate was estimated at the time to be over five thousand pounds. His second wife, Anna Maria Fielding's grandmother, is accused by Osborne of having ruined George Carr by her extravagance:

> She induced him to expend large sums on the domain of the family residence, in forming miles of flower beds through the fine old timber lands of the estate; on building grottoes and ornamental temples, forming extensive puzzle walks confined by high hedges of boxwood kept beautifully trimmed and constituting a walk on a mound of a "double ditch" of the estate, down to cliffs on the sea shore where steps were cut in the rock leading to the Graige bath houses. All these improvements, which required large annual outlay to keep in order, encumbered the Graige estate ...[12]

George Carr, to meet his debts, made over a part of the estate *'worth twenty thousand pounds'* to his overseer, and more financial troubles followed. By the time the estate descended to the Osborne family in 1835 there was very little left and Richard Boyse Osborne accuses a relative, Thomas Boyse, of having cheated him out of his inheritance. There is an obvious bitterness in Osborne's recollections, and there is no way of knowing what truth there is in his allegations of extravagance, incompetence, and sharp dealing. It is of interest, however, to note that many of Mrs Hall's recollections of her family home in Bannow accord with Richard's descriptions and a further point of interest is that Mrs Hall obviously did not share Richard's view of Thomas Boyse, because she dedicated the Third Edition of her *Sketches* to him with the inscription 'To Thomas Boyse, Esq. of The Grange, Bannow; These Sketches chiefly relating to a District over which he so worthily and beneficially presides – are Inscribed, with sentiments of sincere respect and esteem, by the author'.[13] Thomas Boyse is referred to by a local Wexford historian, Father Butler, O.S.A. as 'a good landlord for his tenants ... and liberal in his political views'.[14] Thomas Moore, the poet, was a guest of Boyse's at Bannow House and has left a description of his visit there in 1835. He found Boyse to be 'a well-informed, off-hand gentleman-like person'. Moore, then at the height of his popularity, was greeted with great enthusiasm by the people of Bannow, who cheered the poet's procession on a triumphal car, accompanied by horsemen, carriages, banners, amateur musicians and nine Muses (to one of whom Moore took a great fancy). There were speeches in front of Bannow House and Moore observed that 'Boyse was very eloquent and evidently in high favour with the people'. Later, the party visited Graige House, Mr Boyse's new acquisition and Moore admired the 'extensive pleasure grounds and the walk to the sea – a sort of garden walk'.[15] Boyse, one may assume, was the model, along with Grogan Morgan of Johnstown Castle (to whom another edition of the *Sketches* is dedicated) of the good resident landlord so often celebrated by Mrs Hall in her stories and sketches.This then, was Mrs Hall's Irish background which gave to her work that 'authenticity' which was so prized by critics and on which her reputation as an interpreter of Irish life was based. Given the nature of the relationships between landlord and tenant, master and servant, Protestant and Roman Catholic, Anglo-Irish and native peasant, her 'authenticity' could be only partial, but it was based on several observed and observable facts, and was presented in an affectionate way in a lively and attractive style.

Throughout Mrs Hall's literary career of writing about Ireland this matter of authenticity came up again and again, and was always commented on admiringly by the critics. The reviewer in the *Literary Gazette* praised the First Series of the *Sketches of Irish Character* for being 'thoroughly Irish, with all the vivacity, blarney, blunders, pigs and potatoes of a Bannow cottage duly set forth', yet there was no falsity about it, rather it was 'sketched with all the warmth and simplicity which says more than any preface could do for the actual truth'.[16] 'It is throughout Irish,' was the verdict of the *New Monthly Magazine*, 'and the picture of Irish manners is not exaggerated'[17]. (One must remember, when reading early laudatory articles in the *New Monthly Magazine* about Mrs Hall's work that her husband, Samuel Carter Hall, had an editorial connection with the periodical for some years and that his opinions would have carried some weight and might have given a bias to critical opinions.

However, while the notices and reviews might at times be over-fulsome they never ran counter to general literary critical opinion as expressed at the time.) In his review of the First Series of *Sketches of Irish Character* the critic in the *Eclectic Review*, commenting on the pleasure which a book about a nearby, yet unfamiliar country can give, noted that 'it is in a semi-civilised country like Ireland that we expect to meet with the most picturesque varieties of character'.[18] These varieties were faithfully portrayed by Mrs Hall according to Allan Cunningham in an article in the *Athenaeum* in 1833 on the literary history of England, Ireland and Scotland which stated 'In her Irish stories Mrs Hall excells. Her rustic maidens are copied from the cottage; nothing can be more faithful and lively, nor are her hinds and husbandmen anything inferior; we nowhere see the Irish character more justly or so pleasantly reported. She sees nature in proper dimensions; there is fancy, but no exaggerations and life always'.[19] The critic in *Fraser's Magazine* who reviewed the Second series of the *Sketches* was the only one to sound a discordant note in the chorus of praise for Mrs Hall's early Irish work. He devoted thirteen pages to the book, taking each story in turn exposing the weakness of the plots, and the innate absurdity of some of the situations. He laughed at Mrs Hall's aim of making 'Ireland agreeably and advantageously known to England', for surely everyone was aware of the beauty of Irish scenery, and the charms of Irish people. The tone of the review is heavily facetious and familiar – 'Be not alarmed, Mrs Hall', 'Dearest Mrs Hall', 'Ah, Mrs Hall, Mrs Hall', but in a more serious mode he insinuates that Mrs Hall may have been guilty of plagiarism in her story 'The Rapparee' which had a plot markedly similar to that of Bulwer Lytton's *Paul Clifford*[20]. The waspishness of the piece suggests that it may have been by William Maginn, (who later wrote a most flattering profile of Mrs Hall for *Fraser's* 'Portrait Gallery'), well known for his critical savagery. The fact that Maginn was himself Irish may have influenced his view of Mrs Hall's portrait of Ireland and he remained the exception among reviewers.

Mrs Hall's 'warmth, sincerity, enthusiasm and 'Irishness' so applauded by the critic who reviewed her children's book *Chronicles of a Schoolroom*[21] in the *Edinburgh Literary Journal* in 1830,[22] was also emphasised by the reviewer of *Tales of Woman's Trials*[23] in the *Literary Gazette* in 1834. Several of these stories had an Irish background, and the critic, identifying some of the qualities which made Mrs Hall so acceptable a writer – 'evident good feeling', 'female purity of ideas', 'occasional sweet touches of nature and neat traits of character' and 'an elevated religious tendency' –

concluded that her talent was 'chiefly shown by her unenforced recollection and description of Irish manners'.[24] *Lights and Shadows of Irish Life*,[25] a three-volume work published in 1838, and based in part on several trips to Ireland that Mrs Hall had made with her husband, confirmed the accepted view that she was a reliable observer of Irish life. *The Sunday Times*, in a review of the dramatic version of one of the tales – 'The Groves of Blarney', playing in the Adelphi Theatre in London, and starring the Irish actor Tyrone Power – while not very enthusiastic about the production, nevertheless pointed out that they had 'been at all times pleased with Mrs Hall's *Sketches of Irish Character*. There is 'a truth, a fidelity and withal a generosity' in her descriptions of 'that libelled land', but the dramatic version lacked 'much of the finish and still more of the ease that has marked her pen'.[26] A week later the book itself was reviewed in the same paper and this time Mrs Hall was praised not only for her 'rich comic humour' and her 'appealing simplicity' that was 'wholly devoid of ridiculous pretension' but above all, for her 'close observation of the national character of her countrymen and countrywomen'.[27] *The Observer*, in its review of *Lights and Shadows* spoke of Mrs Hall's 'adherence to truth, which is in this case, but another name for nature'. The 'blending of variety, humour and pathos' are 'altogether illustrative of Irish life,[28] and must appeal to every taste. *The Spectator*, however, in its review of the story 'The Groves of Blarney' mentions the darker side of Irish life, with which Mrs Hall is obviously familiar – the instability, impetuosity, and fondness for alcohol which are 'national characteristics' as are 'crimes of violence equally Irish'.[29] *John Bull* nodded approvingly at this delineation of the national Irish character, citing it as proof that while Mrs Hall was 'a national writer … [she was] not therefore a prejudiced one'.[30] (The *John Bull* reviews of Mrs Hall's work must at this stage be treated with some caution as Mr Hall had moved from the *New Monthly Magazine* and was now sub-editor of *John Bull* for a year or so.) The *Weekly Dispatch* acknowledged that Mrs Hall was a trustworthy reporter on Irish life and character, but regretted the levity it detected in the stories. Unlike Maria Edgeworth she made no attempt to humanize the Irish but, said the reviewer, 'travels through Ireland, and although she describes all the inferiorities that she witnesses, she attempts to joke them away. This is mistaken kindness to Ireland. It is nationality of the best sort of heart, but the very worst to the head'. The reviewer concludes sourly that 'the world is getting sick of Irish tales'.[31] Fortunately for Mrs Hall and the other Irish writers of her day, that time had not yet come.

The *Weekly Dispatch* review is unfair to Mrs Hall, when it accuses her of levity, and by implication, lack of concern about the moral, physical and social condition of the Irish people. She did care, very deeply, as we shall see, and tried by her writings to improve the manners and morals as well as the physical well-being of those she called her countrymen and countrywomen. *Stories of the Irish Peasantry*,[32] a collection of little tales and sketches that had appeared in *Chambers's Journal* throughout 1839 and early 1840 was published in book form in June 1840. The stories illustrated Irish peasant faults and were avowedly written to help humble readers cure these faults. In *Marian: Or a Young Maid's Fortunes*,[33] a novel set in England, and published in 1840, the character most singled out for praise by critics was an Irishwoman, a servant called Katty Macane in whom, according to the reviewer in the *Monthly Chronicle*, 'the devotedness, sagacity and rich humour of her country are felicitously displayed'.[34] The *Athenaeum* reviewer saw this character as an 'Irish cook done to the life ... shrewd, unprincipled, warm hearted,[35] and the *New Monthly Magazine*'s critic found Katty Macane (variously described by reviewers as a cook, a nurse and a washerwoman) to be 'the real heroine of the new novel' and opined that Mrs Hall had 'never before been so successful in delineating the Irish character'.[36] The critic in *The Sunday Times*, reviewing a later edition of *Marian* ('The Parlour Library' series) in 1847, also concentrated on the portrayal of the Irish servant – 'the best character in the story ... who discourses with all the volubility and eloquence of her nation as often as she is introduced to our notice'. Mrs Hall knows 'the peculiarities of the Irish people' and 'has studied their characters – she understands traits of feeling and can appreciate acts which appear all but destitute of meaning to the common observer'; indeed 'her delineations of the Irish peasantry in most instances are distinguished for their truth and accuracy'.[37]

By the time that review appeared, two significant works by Mrs Hall had already been published. The first, *Ireland, its Scenery, Character, etc.*,[38] had been written in conjunction with her husband, Samuel Carter Hall, and been published in 1841–43 to great acclaim, while the second, *The Whiteboy: A Story of Ireland in 1822*,[39] a full-length Irish novel from her own pen, had appeared in 1845 and had received generally favourable reviews. The guide book, *Halls' Ireland*, earned for both the Halls widespread praise and was reviewed in over sixty periodicals in England and Scotland, an indication not only of Mrs Hall's popularity as a writer, but of the trust placed in her as a faithful reporter on Irish life. It was Mr Hall who provided the facts and figures in the guide-book, but it was Mrs Hall's little stories (dismissed by *Fraser's Magazine* as 'nouvellettes'[40]) that gave it life, or, as *The Sunday Times* reviewer said, 'infused soul' into the book.[41] *A Week at Killarney*, one of the part works that made up the book evoked special praise from *The Sunday Times*, which declared that her description of a deaf and dumb boy 'was worth all Sterne ever wrote', for '*its* feeling is spontaneous and genuine' and that Ireland should be grateful to her for she had 'done as much for that ill-fated land in *prose* as Moore, the bard of all time, has achieved in verses'.[42]

The Whiteboy was not universally praised, although Mrs Hall's talents in depicting Irish life and character were never in dispute. The critic in the *Gentleman's Magazine* was impressed by her knowledge of the Irish people and by her interest in her welfare and compared her 'lively, spirited and accurate descriptions' to those of Miss Edgeworth, but he was unhappy with the 'morbid liberalism' in the story.[43] In *John Bull* a reviewer of the book thought Mrs Hall far superior to the general run of Irish story tellers yet he did not approve of her aims – those of 'inculcating political truths, and exhibiting the consequences of a faulty legislation in what regards "justice to Ireland"'. As far as he was concerned these were the 'least valuable portions of the book'. Mrs Hall was 'not the person destined to solve the knotty problem of giving peace and contentment to her country ...' Better that she should stick to what she was so good at, delineating 'the manners, feelings, superstitions and sympathies of the Irish character'.[44] The *Atlas* had no objection to Mrs Hall's aspirations, and although the evil of absenteeism, which occupied the principal part of the story had already been 'so well exposed in Miss Edgeworth's tales in *Valentine McClutchy* [sic] and in many of Mr Lever's works, yet Mrs Hall had managed to 'form an interesting story out of old material'. Indeed 'any story, relating to Ireland, from the pen of Mrs Hall is certain to be amusing and instructive ... from that lady's knowledge of Irish character'.[45]

Another full-length novel by Mrs Hall, *Midsummer Eve*,[46] was published in book form in 1848, having already been serialised in the *Art Union Journal* the publication edited by Samuel Carter Hall but although it was for the most part set in Ireland, there was no critical recognition of its having any distinctively Irish flavour. *Midsummer Eve*, subtitled *A Fairy Tale of Love*, was an airy piece of fantasy which *The Observer* found to be 'a homely but beautiful story in every respect'[47] and the *Atlas* described it as 'a fairy tale upon which Mrs Hall has raised a dramatic structure coloured throughout with a more political spirit than we usually find in her stories'.[48] Nothing of Irish interest there, and the 'politics' referred to were those of the art world. Mrs Hall's last long story of Irish interest, *The*

Fight of Faith,⁴⁹ published in 1869, was a novel set largely in seventeenth-century Ireland. It is a rancorous attack on Papism and the Church of Rome, and displays an unexpected and unpleasant bigotry that was generally absent in her earlier work. Indeed, her religious tolerance was one of her attributes admired by critics.

It was Mrs Hall's good fortune that she wrote her Irish stories and sketches at a time when there was a deep and genuine English and Scottish interest in literary work out of Ireland. From the early years of the nineteenth century, and for four decades, Irish writers were very much in vogue. Maria Edgeworth, Lady Morgan, Eyre Evans Crowe, 'The O'Hara Brothers' (John and Michael Banim), Gerald Griffin, George Brittaine, Thomas Crofton Croker, William Carleton, Joseph le Fanu, Thomas Moore, Charles Maturin, all in their turn sold well in England and Scotland and were accorded serious critical attention. Reviews of their works (and those of lesser authors) took up a great deal of space in the literary journals and their merits and demerits were vigorously discussed. At times the criticisms were severe and a note of *ennui* can sometimes be detected, as in the *Fraser's Magazine* 1830 review of William Carleton's *Traits and Stories of the Irish Peasantry*, a review which managed to attack Thomas Moore on the ground that 'his nationality was debatable', the Banims as being 'offensively vulgar', Crowe as being 'sometimes too refined and too metaphysical' and Griffin as 'occasionally tedious and un-connected'. Lady Morgan's 'absurdities' were 'unquestionable' and only Miss Edgeworth was spared the reviewer's scorn, for she stood alone as being 'classically Irish'.⁵⁰ Nevertheless, most reviewers and ordinary readers would have agreed in those years with the comment in the *Edinburgh Review's* comprehensive critique in 1826 of Irish novels that 'at present Ireland bids fair to be the great mart of fiction.'⁵¹

It continued to be so for another two decades and more, and the emergence of a new talent in 1830, that of William Carleton, a true chronicler of the Irish peasant society from which he had sprung, sustained English and Scottish interest in Irish literature. By the end of the 1840s, however, the wares in the mart were losing their appeal, due to a number of factors, not least among them the horror of the famine of 1845–49 which made fictional offerings from Ireland look tawdry when seen beside factual reports. Furthermore, a new generation of English authors had appeared on the scene, and it was not surprising that writers who included Dickens, Trollope, Thackeray and the Brontës should eclipse Irish authors in popularity. The one Irish author who retained his favour with the

English reading public was Charles Lever, but his is a special case. He could fairly claim to be more international than his Irish contemporaries, not only in the sense that many of his novels had foreign settings, but because he was more outward-looking and more aware of a European perspective.

While it lasted, Mrs Hall profited by the boom in Irish books. Her stories and sketches were unlike those of other Irish authors, and although they contained plenty of violent incident, the overall impression left on the reader was that of a very beautiful country, favoured by nature, where the people were basically warm-hearted. They had faults, it was true, but these could be cured, and the Irish peasant character could be improved if the people were treated by their superiors, both English and Irish, in the right way, kindly, but firmly. Mrs Hall's 'feminine delicacy', a description beloved of reviewers, conveyed an impression of a gentle domesticity that could be found even in the wild land of Ireland. As a female, her name was often linked by critics with those of Miss Edgeworth and Lady Morgan but reviewers differentiated between the works of the three authors. The *Edinburgh Literary Gazette* in its review of the first series of the *Sketches* marvelled that Mrs Hall was able to follow in the footsteps of Miss Edgeworth and Lady Morgan who had already reaped a 'full harvest' in the Irish field and yet procure some gleanings of her own: 'Mrs S.C. Hall has had the sense to make a further essay, and these two volumes prove that all which is rich and rare has not been picked up by those who have gone before her'.[52] The *Monthly Magazine* was more enthusiastic in its review of the *Sketches* and stated boldly: 'Miss Edgeworth's place is adequately supplied, and her indolence no longer to be regretted'. Mrs Hall's command of the 'native idiom' was 'more thoroughly complete ... than any of her scribbling contemporaries'.[53] The *Edinburgh Literary Journal* rated her worthy to take her place with her highly talented countrywomen, whose names are linked with [Ireland's] literature' and was pleased to see that the work 'lacked the formality of instruction' found 'in 'Leadbetter's' [sic] *Dialogues*, and even in Miss Edgeworth's writings'.[54] However, in the review of the Second Series of *Sketches*, in 1831, the critic in the same periodical put it rather differently: 'Mrs Hall is a writer after our own heart. If she does not possess Miss Edgeworth's masculine power of scanning character, she at least unites to the benevolent and tempered utilitarianism of the author of *Ennui*, more feminine gentleness, and all a woman's intuitive knowledge of the workings of the human heart'. She was to be preferred to Lady Morgan, for although she possessed that writer's 'sentiment and imagination' it was 'untainted by her ladyship's false and obtrusive philosophy'. Yet she was 'less intensely powerful than either of those gifted females' although she united 'in a high degree the good qualities of both her countrywomen, tempered in her, more than in either of the others, with all a woman's mildness'.[55] (Obviously a less challenging female, as far as male reviewers were concerned.)

The *Athenaeum* critic who reviewed the second series of *Sketches* also made the comparison between Mrs Hall and the other two popular Irish women writers, and had no doubts about where she stood in relation to them: 'If she does not possess the exquisite fineness of observation and finish of touch displayed by Maria Edgeworth in her national tales – if she manage the farce of brogue and humour with less *gusto* and buoyancy than Lady Morgan – she stands in the chapter of female Irish novelists next to those two accomplished women'.[56] The comparisons with Miss Edgeworth and Lady Morgan continued right up to the end of Mrs Hall's career as an Irish writer, marked as it was by the publication of *The Whiteboy* in 1845. The *Gentleman's Magazine* reviewing the novel, paid tribute to Mrs Hall for the valuable information she always gave about Ireland, and doubted whether it could be found in the work of any other Irish writer, except 'in some of Miss Edgeworth's Irish tales'[57], and the *New Monthly Magazine* (with which Mr Hall was no longer associated), in a thoughtful review of *The Whiteboy*, spoke of Mrs Hall as 'undertaking at once the duties of Miss Edgeworth and of Lady Morgan'[58] in putting forward the case for the consideration of Ireland's woes.

Mrs Hall's cultivated 'Irishness' was not the only factor in her success in winning over English and Scottish readers, although it was a very large one. She was a woman writer at a time when women writers were numerous, popular and prolific. The literary merits of the women writers who followed in the wake of Jane Austen and Fanny Burney are not always readily discernible, nor have all of them retained their popularity to the present day, but in their own time writers such as Mrs Barbauld, Felicia Hemans, Catherine Gore, Letitia Landon, Amelia Opie and Grace Webster were well known and highly regarded. They all benefited from the new fashion for periodicals, ranging from the small and obscure to the large and famous, and Mrs Hall was no exception. Furthermore, she was very much in tune with the taste of the times for amassing knowledge and for being lectured. It was an age of didacticism and Mrs Hall was nothing if not didactic. Her peculiar appeal lay in the skill with which she combined her Irishness with her desire to teach, and the palatable mixture of entertainment and education which she offered to a receptive public.

TWO

Mrs Hall – Marriage and Markets

Samuel Carter Hall, the young man whom Anna Maria Fielding married on the 20th September 1824, had been, like her, born in Ireland. His father was a military man from Topsham, near Exeter in Devon and had been sent to Ireland with his regiment, the Devon and Cornwall Fencibles, in 1795. Colonel Hall was stationed at Geneva Barracks in County Waterford, and it was there that Samuel Carter, the fourth of twelve children, was born in May, 1800. During the Rebellion of 1798 the Colonel's regiment was in charge of maintaining the peace in County Kerry, and according to his son, there were no violent incidents there. Unfortunately for the Hall family, the Colonel caught the mining fever that was raging at the time, and did not return to England with his regiment when it went back in 1802, but put his money into the development of copper mines in Kerry and West Cork. The copper was there, all right, and the mines were profitable for a time, giving much-needed local employment but they eventually failed. The abandoned mine workings can still be seen and have the melancholy air of all doomed ventures. Mrs Hall, the Colonel's wife, who must have been a woman of spirit, opened a shop in Cork and somehow the family survived. In his book, *Retrospect of a Long Life, from 1815 to 1883*,[1] which is the only source we have for this family information, S.C. Hall speaks lovingly and admiringly of his mother, adding that she died of typhus fever contracted while visiting the poor. Colonel Hall lived in Chelsea until his death in 1836, and his son raised a monument to him in Kensal Green Cemetery.

Young Samuel Carter left Cork in 1821 and got a job in London as a literary secretary to the exiled Italian poet, Ugo Foscolo. In his memoirs he hints at temptations that lay in wait for him in those days and in that company:

In the year 1882, I pause for a moment, to make record of my thankfulness to God, who, in 1822, preserved me from my first great peril – to mind, heart, and soul. I do not refer ... to escape from the sirens, though that is cause for gratitude, but to the Mercy that saved me from the taint of infidelity, to which I was, then and there, more than merely exposed. I do not mean that Foscolo strove to corrupt me; but he assuredly placed me in the way of strong temptations.[2]

However, Hall *was* saved, because it was at that time that he met his future wife, and was then 'in no danger from that ever-potent source of danger to youth'. Miss Fielding had been living in London with her mother and step-grandfather, George Carr, since 1815, and was then 23, quite an advanced age for those days. Mr Hall records that when George Carr died in 1823 he left no will, and that, consequently, Anna Maria, although she was his adopted daughter, got no money. Richard Boyse Osborne, the grand-nephew of George Carr, tells a different story. In his *Diary* he states that George Carr settled an annuity of £100 on Mrs Hall, to be given as a wedding present. This was paid to her throughout her life, and after her death in 1881 passed on to her husband.[3] If this was so, it would have meant a great deal to the young couple. Samuel Carter had left the employ of Foscolo and was working as a parliamentary reporter and as what we would now call a freelance journalist when he met Anna Maria Fielding. A few days after the wedding he received a payment of £40 for some literary work, and this, he says, was enough for 'the church fees and a wedding trip to Petersham, near Richmond'.[4] Enough money came in that year for the Halls to make a visit to Bannow in 1825, the first of many Irish trips they were to make together.

Samuel Carter Hall's journalistic career was varied, but not brilliant, and he seems to have inherited his father's bad luck. His first publishing venture, a periodical called the *Literary Observer*, collapsed in 1823 after only six months. From 1826 to 1827 Hall edited a little publication, the *Spirit and Manners of the Age*, and he worked for a time between 1829 and 1830 on the *Morning Journal*. In 1830 he became sub-editor of the *New Monthly Magazine* and rapidly succeeded Thomas Campbell as editor. In 1831, however, Hall was replaced by Bulwer Lytton and demoted to sub-editor. By the following year he was back in the editorial chair where he stayed until 1836 when he was displaced to make way for Theodore Hook. In this game of musical chairs he moved in 1837 to *John Bull*, the magazine which Hook had edited, and became sub-editor there. Hall also set up a magazine called the *Town*, which he edited for a year, and for which he received £1,000; he failed in that venture too. On then, as general manager of the periodical, *Brittania* in 1839, while at the same time he wrote leading articles for the *Watchman*, a Methodist newspaper. It must all have been very unsettling, not only for him, but for Mrs Hall, who seems to have been the very model of a devoted wife. The worst blow to the couple came in 1837 when the *Amulet*, a little periodical which Mr Hall had founded in 1826 and had been editing ever since, collapsed with the crash of its publishers, Westley and Davis. Hall was liable financially, and lost a great deal of money. Then in 1839 he was retained by the print publishers, Hodgson and Graves, to edit the *Art Union Monthly Journal*. At last there was some continuity in his journalistic career, for he remained editor of the periodical, renamed in 1849 the *Art Journal*, until his retirement in 1880. The periodical was popular and influential, and Samuel Carter Hall is credited as having exposed the trade in fake 'Old Masters'. Yet even this period of employment was not without its strains and stresses. Hall bought a major share in the *Art Union Monthly Journal* but, according to his memoirs, it was nine years before the periodical paid its way, and then, only three years later, in 1851, he was forced by lack of money to sell his share and revert to being a paid editor at a salary of £600 per annum. He also got involved in two libel actions, one in Warwick in 1855, (which the *Journal* lost), another (which was settled out of court) in 1878.

On reading this resumé of Samuel Carter Hall's journalistic career one might reasonably conclude that Anna Maria Fielding made a bad match, materially speaking, because it was a marriage beset by financial worries and marked with professional disappointments. Mrs Hall's pension of £100 per annum from the Civil List, awarded in December of 1868, would have eased the financial strain somewhat, but it is significant that the golden wedding gift given to the couple by their friends in 1874 was in the form of an annuity purchased with the money that was collected for that purpose. Mr Hall was himself awarded a Civil List pension in April, 1880. On the other hand, marriage to a man who was successively editor (or sub-editor) of three periodicals, was Mrs Hall's entrée to the literary world. Her first Irish story, 'Master Ben', a fond recollection of the Wexford schoolteacher who had given her her early lessons while she lived with her grandmother in Bannow, was published in the *Spirit and Manners of the Age*, the monthly periodical. It had been edited in its first year, 1826–27, by Samuel Carter Hall, who presumably still had some influence with the publishers Westley and Davis, who were also the publishers of the *Amulet*, the journal of which Hall was editor from 1827 until its collapse in 1837. The *Spirit and Manners of the Age* had started life as a weekly and its first issue was dedicated to William Wilberforce. In keeping with the spirit of the age the periodical offered readers 'much useful information on the subject of foreign countries, particularly as it regards their moral and religious condition', and promised to be 'an unexceptionable source of amusement and instruction.[5] By 1827, however, the periodical had become a monthly, rather than a weekly, but the high moral tone persisted, as did the interest in foreign customs, exemplified by such articles as 'An Illustration of Persian Manners', a description of the lives of 'Hindoo Fishermen' and 'Letters on the Moral and Religious State of South America'. Mrs Hall's Irish story fitted neatly into the magazine's formula, and in its simplicity seemed to have much in common with Miss Mitford's little essay, 'A November Walk' also published in January 1829.

Mrs Hall, in what she called 'A Rambling Introduction' to the Fifth Edition of *Sketches of Irish Character*, in 1854 recalled the day she wrote her first story:

I can remember how my voice trembled, when, little more than a bride, I ventured to read to my husband the sketch of "Master Ben" ... glancing from the seamed and blotted page, to his young face, then unmarked by the cares and anxieties he has since often felt both for himself and me, and wondering if it were really true that I – *I* could write anything worth being printed – and paid for.[6]

Her husband, as editor, gave his verdict that yes, she could write, and be paid for it, so 'Master Ben' was followed by four more sketches that were published in the *Spirit and Manners of the Age*, all based on childhood memories. Two of these 'sketches' should be more accurately described as stories, because 'Black Dennis' and 'Mary Clavery's Story' are both more than simple reminiscences. They are highly-coloured tales of tragedy and vengeance with enough of drama and suspense to catch the imagination of readers, and would have enlivened the rather staid pages of the little journal in which they appeared. The publishers, Westley and Davis, evidently thought so because they made Mrs Hall an offer that astonished her: 'When a publisher offered me a hundred pounds if I would write a new Sketch and collect those I had written into one little volume for publication, honestly I told him I thought it too much, though I did certainly exult in the "great fact;" and then how astonished I was to find myself famous in the generous pages of the "*Literary Gazette*"!'[7] The collected sketches, and several new ones, appeared in April 1829, and Mrs Hall did indeed become "famous", and her career was well launched. The *Spirit and Manners of the Age* reviewed *Sketches of Irish Character* in its May issue, and found it full of 'freshness, originality and vigour'. It was based on 'actual personal observation' and although it had 'no pretensions to depth' it conveyed 'details of human nature always just and often far from superficial'.[8] That was the official editorial view, but the personal, domestic reaction was rather different. Mr Hall, according to his wife writing in 1854, was worried about the shape their marriage might take if she devoted herself to writing:

> I can also remember, how fearful my husband was that literature – its cares, its claims, and its fame – would unfit me for the duties which every woman is bound to consider only next to those she owes her Maker. I daresay I was a little puffed up at first, but happily for myself, and for those who had near and dear claims upon my love and labour, I very soon held my respon-sibilities as an author second to my duties as a woman; they *'dovetailed'* charmingly, and I have never found the necessary change to domestic from literary care, though sometimes laborious, not only heartful, but pleasant.[9]

It is tempting to look on Mr Hall's fears as being a mixture of selfishness and jealousy, and indeed it is hard to see them in any other light. A wife who devoted herself to literature, and made a career out of it as many of her contemporaries did, would not be able to give him her single-minded devotion and attention – the household might not be properly run and he would suffer a thousand little discomforts. Then, too, a wife who was a famous literary figure would overshadow him, had indeed already done so. He had graciously read her little sketches, had encouraged her, and had helped her have her work published, condescending to her from his position of professional authority, and all at once she had achieved a greater measure of success within a few months than he had achieved in years. Small wonder he was 'fearful', but the couple obviously worked matters out between them and she managed, somehow, to salvage his pride. She never forgot the incident, it seems, if, after more than twenty years she could describe it so vividly, and even very late in life we hear an echo from the past. In Mrs Hall's story 'Building a House with a Tea-cup' (which was accompanied by another, 'Digging a Grave with a Wine Glass') published in 1875 in a collection of temperance tracts called *Boons and Blessings*, a wife admits: '"I was not nearly as careful in my early married life as I should have been of his [husband's] little home comforts."'[10]

In the Hall household there were very few to lay claim to Mrs Hall's 'love and labour', apart from Mr Hall, and her own mother, Mrs Fielding, who lived with the couple until her death, aged 83, in January, 1856, for there were no children of the marriage. None surviving, that is. In a footnote to his book of memoirs, *Retrospect of a Long Life*, S.C. Hall mentions the monument which he erected in 1836 in Kensal Green, to the memory of his father, Colonel Hall, and he adds that also inscribed on the stone (along with the name of a faithful servant, Hannah Davey, who had been with the family for 50 years) is the name of a child, Maria Louisa. This, he says, was 'the only child we had who lived, and her life on earth was very brief'.[11] From this, one gathers that Mrs Hall had suffered miscarriages, or stillbirths, or both, and that Maria Louisa died in infancy. There is no other reference to a child, or to repeated disappointments in Mr Hall's memoirs, but in another work, *The Use of Spiritualism* written in 1884, Mr Hall is more informative. Both he and Mrs Hall, although devout Christians, believed that messages could be sent from beyond the grave and that contact could be made by the living with the dead – to them spiritualism was 'reasonably, rationally and scripturally true' and they attended many séances. In his book of memoirs, *Incidents in My Life*, Daniel D. Home, the celebrated American medium, mentions the Halls' presence at one of the séances he conducted during his London visit in 1863. What happened does not seem to be either interesting or significant, but for Home it was compelling proof of his powers. Mrs Hall, as witness, 'deposed to having received the present of a lace cap from the deceased Mrs Home, laid by supernatural hands on her knee'.[12] There is also a reference to a séance held in the Halls' house at Campden Hill in March, 1866; 'Five persons assembled [there] Mr and Mrs S.C. Hall, Lady Dunsany, Mrs Henry Senior and Mr D.D. Home. Our host and hostess said repeatedly to each other during the evening "We have never had anything like this before," and they have certainly seen more wonders in spiritualism than most people'.[13] After Mrs Hall's death in 1881 Mr Hall made many attempts to get in touch with her, and finally, he relates in *The Use of Spiritualism*, on 6th January, 1883, her birthday, he heard her voice, '"This is my birthday, and I rejoice to come. Our child is here"'. On the 30th of the same month he heard a longer message: '"If you could have seen me when my spirit took its immortal form, you would have said 'Thank God my Marie is free from all suffering – united to her child and mother, her relations and friends – how happy she is' ... . My pet bird is here and my little Blackie, and my child – our child – Carter; and when are you not with me?"' Mr Hall comments, 'God gave us but one child, who lived but ten days in this life. She now (I believe invariably) comes to me with her mother who tells me '"Our daughter is a young maiden ... she has blue eyes and hair a little lighter than my own"'.[14]

Mrs Hall has left no record of what must have been times of great sorrow for her. Even in an era when infant mortality was high and parents were reconciled to the probable loss of one or more children it was a cruel fate to have been left with no child at all, especially if she had carried several to full term. She had gone through the hazardous experience of childbirth and had suffered its dangers but had received none of the rewards. An added torment would have been the belief that she alone was responsible for the failure to bring healthy children into the world. She had failed in her primary duty as a woman, and her success as a writer could be no more than second-best. This belief may very well have lain at the heart of her relationship with her husband, whom she was always concerned to portray as the superior partner in the marriage. That was, of course, the custom and convention of the times, regardless of what the reality might have been, but in Mrs Hall's case – she who had failed as a woman – there was a sound personal reason for her conviction. Whatever her private fears and doubts, and whatever her sorrows, she seems to have found consolation in her strong religious faith, and to have reconciled herself to God's will. In the only direct reference to her childless state that I have been able to find, she says, 'This terrible anxiety about my friends' children sometimes makes me thankful that God recalled those he lent me for a time; – it is so much easier to dictate than to do – to preach rather than practice'.[15] This statement is an authorial intrusion into a story called 'The Naughty Boy', which is all about the importance of firm moral training in the upbringing of children. Many of her stories for and about children emphasise this aspect of child-rearing – much more important in her view than secular education. In the dedication to a little Irish girl, Lizzy Grogan Morgan of Johnstown Castle in Wexford, of a children's book, *The Juvenile Budget*, in 1837, Mrs Hall states 'I love little children dearly, and I never wrote a tale for their amusement, without thinking of their improvement, and endeavouring to promote it, by all the means in my power. It is one of the many blessings of life, dear Lizzy, that the performance of our duties increases our happiness'.[16] So, she came to terms with her private sorrow, and sublimated it in the way she knew best – writing improving and uplifting tales not only for children, but also for adults, especially Irish peasants, as we shall see. At home she concentrated her energies on making life comfortable for her husband and her mother, whom she loved dearly. As she said in a letter to a friend, a Mrs Lover, 'My home affections have been very, very few – but they are painfully strong.'[17]

Domestic duties for Mrs Hall did not include floor scrubbing, polishing, washing, ironing and rough work, or indeed, cooking, because in common with most women of the middle class in those years she employed servants. She was the overseer and the director, and also observer of the behaviour of those she employed. They provided the raw material of some of her stories, as she admits in the Introduction to the Fifth Edition of *Sketches*, quoted earlier:

Having a tolerably good Irish cook, I was provoked by her silence – a very unusual fault – I could never get her to answer a question, and when I sent for her she would hold the door in her hand, and after a brief "Yes, ma'am"; or "No, ma'am;" start off with such evident delight at her escape, that I resolved to know why she desired to avoid me.

"Mary", I said, one morning, "I want to know why you answer my questions so abruptly. Shut the door – there, let go the handle, and come here – have you any fault to find with your situation?"

"No, ma'am – never a fault at all with the place – Oh, no! God bless you."

"Then what is the reason that whenever I send for you I can hardly get you to answer my questions? And when I go into the kitchen, you invariably rush into the scullery?"

"It's just nothing particular, ma'am."

"But there must be a reason for such un-servant-like conduct."

"No reason in life, ma'am."

Mary had got back to the door, and regained firm possession of the handle.

"I know you can talk fast enough sometimes, and yet, as I have said before, you will not even answer my questions." She made a sudden rush out – then returned – thrust her head through the opening, and exclaimed piteously – "Ah, then, let me alone, ma'am, dear, you know you'll be putting me in a book!"[18]

One sympathises with the cook, one of the 100,000 general domestic servants in London who worked for eighty hours per week, earning an average of ½d per hour and their keep. Mrs Hall put this reluctance to appear in print as an Irish superstition, but it was a natural desire to guard one's privacy. Of course, the idea that a servant, especially an Irish one, would have sensibilities, would never have entered Mrs Hall's head any more than it would that of any contemporary employer. In her sketches and stories she often extolled the virtues of Irish servants but concentrated on their fidelity and loyalty to their employers in whose shadow they lived. Her view of servants in general is expressed in the book *the Juvenile Budget*, mentioned earlier. 'Servants', she said, 'because they have not the advantages of good education, are often narrow-minded; and while you remember to treat them with civility and kindness you must avoid conversing with them, as their general sentiments *can* do you no good, and *may* do you much harm.'[19] It was quite in order for Mrs Hall to engage her servants in conversation and listen to their quaint sayings which could be used to comic effect in her little stories, but it was another matter to take their views seriously. This attitude may jar on the modern reader, but we always have to remember that we are talking about a vanished world, where society was stratified more firmly than it is today, and where most members of the lower classes *were* uneducated and consequently, as Mrs Hall pointed out, were narrow-minded and ignorant. It was sensible of her to insist that young children should not be put into the exclusive care of such people, or even spend too much time in their company. Mr Hall, in his book of memoirs, states that Mrs Hall had made some 'wise and useful' observations on the 'important subject' of household servants. He quotes them extensively, but they may be summed up in one paragraph:

Let your servants be treated as part of your family; see to their comforts as they see to yours; lessen their wants, and be sure that *your* wants will be less and less. There is a "familiarity that breeds contempt". But if we can be "familiar, yet by no means vulgar," we can be so without sacrificing an iota of dignity. The servant who presumes upon it is one to get rid of as certainly as she would be if you knew her to be a thief.[20]

Mrs Hall's literary début in the *Spirit and Manners of the Age* as a writer of Irish stories, was very quickly followed by more appearances in the periodicals. The *Amulet*, Mr Hall's own little journal, printed five stories by her in 1829, and continued to publish her work until the magazine went out of existence in 1837. The *Amulet*, like the *Spirit and Manners of the Age*, had high moral aims. These were spelled out in the Preface to the January edition in 1827: 'The especial object of the *Amulet* is to blend religious instruction with literary amusement; so that every article it contains shall bear either directly or indirectly, some moral lesson which may impress itself strongly on the mind by means of the pleasing language and interesting form in which it is conveyed.'[21] During its lifetime the magazine published the work of many of the best and most popular writers of the day, including John Clare, Miss Mitford, Samuel Taylor Coleridge, Maria Edgeworth, Thomas Hood and William and Mary Howitt. Mr S.C. Hall, whatever his own shortcomings as a writer (and there were many), was a good editor, quick to spot the talent of others.

Now, here we come to a discrepancy between Mrs Hall's

recollection of the writing and publishing of her first story, and the recorded facts. I have already quoted the passage from her Introduction to the Fifth Edition of *Sketches of Irish Character* in 1854, in which she recalled that 'Master Ben' was her first published story. Not so, because in 1826, a short tale, or tract, 'The Murmurer Instructed' appeared in the *Amulet*, signed 'A.M.H. ...' Not only are the initials hers, but it bears all the signs of her style, as does another story 'The Gipsey Girl', also signed "A.M.H. ..." and published in the *Amulet* in 1827. The latter story was highly praised in the *Lady's Magazine* in its review of the *Amulet* for the year 1828, and was reprinted in its entirety. No author's name was given, but the reviewer opined that 'of the prose pieces, the 'Gipsey Girl' is among the best'.[22] In another review of the *Amulet* in the same magazine in the following year, reference was made to 'the editor and his ingenious lady'[23], proof that Mrs Hall's literary strivings were already known in the world of journalism. Indeed, by the end of 1828, she became an editor in her own right, having produced the Christmas Annual, The *Juvenile Forget-me-Not*. The reviewer in the *Lady's Magazine* found this a charming little volume and paid tribute to Mrs Hall 'who has adorned it with some of her own compositions; The 'Star'; The 'Young Rebel', and 'The Savoyards'. All these things are creditable to her talents and indicative of her good feelings.'[24] The story 'The Savoyards', a sad and pious little tale about the wretched youngsters from Savoy, who performed on the streets with an organ-grinder, and sometimes a monkey, was often reprinted in later years, as was 'The Gipsey Girl'. It is an early indication of Mrs Hall's deep and genuine interest in and compassion for the poor, the suffering and the exploited, that she should base a story on these little creatures who had been driven from their homes in France by abject poverty.

London in the nineteenth century was no earthly paradise, but it drew to its stinking streets those from far and near who believed that it held the promise of a better life. The city had grown rapidly in the latter half of the eighteenth century, reaching a population of one million by the time of George IV's coronation in 1821, and the growth rate accelerated so fast that the city services could not cope with the increased numbers. The water supply, mostly from public standpipes, was inadequate and impure, coming as it did from the filthy Thames, which was the recipient of London's sewage. Some people paid for piped water from one of the water boards, but the supply was erratic and not always adequate. There was no proper system of sewage – merely 'leaking pipes, uncovered cess-pits, stinking gullies, rotting privies and gas-filled sewers'. It was not only the poor in the over-crowded slums who suffered the ravages of cholera and typhus, although they were the worst hit; the well-to-do, including the Royal family in Buckingham Palace, were also uneasily perched over a potentially lethal underground drainage system. It was not unknown for rats to invade the houses of the rich and fashionable and there were reports of them attacking well-born infants in their nursery cots. It was not until late in Mrs Hall's lifetime that London was provided with a proper water supply and sewerage system – the first Public Health Act was not passed until 1848 and the construction of the main sewer under the new Embankment from Westminster Bridge to Blackfriars Bridge did not take place until 1864–1870. To live long in those days one needed an iron constitution, which obviously Mrs Hall had; possibly inherited from her mother, Mrs Fielding, who lived to be 83. It is ironic that both Mrs Fielding and Mrs Hall lived such long and active lives in the unhealthy atmosphere of London, while Mrs Carr, the grandmother whom Mrs Hall recalled so lovingly, had died at the early age of 54 in the pure, clean air of Wexford. Mrs Hall, of course, for most of her London life lived in one or other of the new suburbs – when she lived in Chelsea in the 1840s it was then a country village – but contagion, especially that of the cholera, was an ever-present menace.

Later in her life Mrs Hall worked hard for charitable causes, but at this early stage, the 1820s and 1830s, she devoted her energies to her career (and to her husband). Her work for the *Amulet* was recognised by no less a magazine than *Blackwoods*, whose reviewer praised the periodical as being in 'every way excellent, free from cant or liberalism in its religion – unobtrusively yet earnestly Christian' and added that Mrs Hall was 'a lady of much taste and feeling, and as need may be, a very lively or a very touching writer'[25]. The *Amulet* was an annual production, originally owned by the publishers Messrs Baynes, who specialised in religious works, and then taken over by Westley and Davis who gave it a slightly more secular tone. This type of Literary Annual or Gift Book was an early nineteenth-century fashion which was introduced to England from Germany by the German lithographer, Richard Ackermann. His first Annual, edited by Frederic Shoberl, was the *Forget-me-Not* which was first published in 1823, and lasted for a full twenty-five years. The fashion caught on, and by 1832 it was estimated that there were 63 different Annuals or Gift Books on sale. Some were ephemeral, but others were longer-lasting. The *Keepsake*, for instance, lasted for thirty years, *Friendship's Offering* lasted twenty-one years, and Heath's *Book of Beauty* for seventeen years.

Shorter-lived, but influential and highly-regarded were publications such as *Finden's Tableaux*, edited by, among others, Miss Mitford, the *Book of Gems*, sub-titled the *Poets and Artists of Great Britain*, edited by Mr S.C. Hall, *the Easter Gift*, edited by L.E.L., (Letitia Landon, a popular writer whose career was ruined by a scandal involving the journalist William Maginn. She later married a Mr McLean and went with him to West Africa where she died under mysterious circumstances – probably of poison), and the *Book of Beauty* edited by Marguerite, Countess of Blessington (whose life was a permanent source of scandal, and whose name was linked by gossip with that of her son-in-law, Count d'Orsay, but who retained her popularity and a small amount of literary fame). The *Amulet*, with its ten-year lifespan, was an honourable inclusion in the list. Its editor, ever-conscious of the proprieties, announced in the preface to the 1832 edition that there would be a change in the contents of the Annual and it would contain 'a larger proportion of articles of permanent interest and value than heretofore, so as to avoid, as far as possible, a very general complaint against the annual works – that they are merely butterflies of a season, and lose their attraction when that season is past'.[26] In fact, compared with earlier issues, the *Amulet* for 1832 is rather feeble. An obsession with death seems to have seized Mr Hall (perhaps understandably, given his family circumstances) and an article on infanticide in the islands of the Pacific, is followed by three poems, the 'Death Cry of Alcestis', 'The Dying Girl to Her Mother', and the 'Death of Eucles'. Then there was an article on the cemeteries of India, and another poem about a dying child. The main illustration is an engraving of a painting by B.R. Haydon entitled 'Death of the First Born'. Mrs Hall's story 'The Mosspits' although it tells the tale of a farm labourer who became involved in a crime of agrarian violence and the suffering this brought on his wife and family, is positively light-hearted by comparison.

As a general rule the Annuals attracted the work of the most talented authors of the day. Frederick Winthrop Faxon, in his study of the Annuals and gift books, lists Byron, Southey, Macaulay, Scott, Wordsworth, Lamb, Coleridge, the Brownings, Dickens and Thackeray among contributors, and adds that 'it was the proper thing to write for these "tokens"'. The volumes were intended as gifts, and 'were such that they might ornament the drawing-room table of the most fastidious without offence either to mind or to eye'. In appearance the Annuals were most attractive and to quote Faxon:

The "really truly" gift books had an appearance and make up all their own, very different from any other volumes. At first they were the small duodecimos [like the *Amulet*], then octavos, and finally some of them appeared as quartos, their bindings were ornate, often to the point of gaudiness. If of leather, the covers were heavily embossed, or profusely gilt, or if a cloth was desired, a watered silk dress gave distinction to the volumes. Some of the earlier ones had engraved paper sides, usually green. Some had bindings with a floral design inlaid with mother-of-pearl. A few had slip-cases to protect their bindings, a ribbon being attached with which to pull forth the volume. Some had large paper editions, sold at a higher price than the regular copies.[27]

There were often flowery borders around the pages of poetry, and the illustrations could number as many as thirty, and consisted of engravings or coloured plates. Engravings were of pictures by, among others, Turner, Prout, Martin, Mulready, Wilkie and Landseer. There was usually also an engraved or coloured "Inscription Plate" on which the donor could write his or her name and that of the recipient. The young John Ruskin was given an Annual by his aunt and not only does he recall it with deep affection, but attributes a deep significance to the gift:

The really most precious and continuous in deep effect upon me, of all gifts to my childhood was from my Croydon aunt, of the *Forget-me-Not* of 1827, with a beautiful engraving in it of Prout's 'Sepulchral' monument at Vienna. Strange, that the true, first impulse to the most refined instincts of my mind should have been given by my totally uneducated but entirely good and right-minded mother's sister.[28]

Ruskin's cousin Charles, son of the 'Croydon aunt' worked as an apprentice in a bookshop, and always brought a book or two home on Sunday for family and visitors to admire. When young John Ruskin was there Charles picked out volumes with plenty of engravings and one of these was truly memorable: 'More magnificent results came of Charles's literary connections through the interest we all took in the embossed and gilded small octavo which [the booksellers] published annually, by title *"Friendship's Offering."*'[29] Although the list of talented authors who contributed to the Annuals and Gift books is very impressive it does not follow that all the contents were of a uniformly high standard. As Faxon says, 'Many of the Annuals in existence diluted their good things with many a mediocre article by unknown or unnamed authors'. Even the well-known authors were often guilty of writing inferior material, a point noted by the reviewer in the *Atlas* in its review of Annuals for 1831. He granted that as

Annuals are such very pretty books, so nicely got up, so beautifully embellished, so neatly bound that it signifies very little whether they be worth reading or not ... . For several years past it has been observed, that the literary departments in these publications have not been well filled. They have boasted of distinguished names, but not of distinguished articles; for, as the best penmen may write a hideous scrawl if they do not take some pains with their writing, so the best authors are capable, as we have seen, of writing very absurd drivel; and nonsense or slop is not the less abominable, because it happens to be written by those who can write better.[30]

The Annuals and Gift Books provided a ready market for Mrs Hall's stories and sketches and reviewers usually greeted her efforts with admiration and respect. She was one of the most popular contributors but she never earned the scorn which the *Atlas* reviewer poured on those who gave less than their best. She was conscientious, and if the plots in her stories were often mechanical, the situations contrived and the dialogue stilted, her reputation was always redeemed by the earnest sincerity which shone through her writing. The stories, absurd though some of them are, never give an impression of having been written to order, or to keep to a deadline or even to make money, but are honest expressions of feeling and belief. Her style of writing was simple and fluent, only occasionally marred by overt and portentous moralising. Her best work was in her Irish stories, but she kept up an acceptably high standard. Now and then she faltered, as the reviewer in *Fraser's Magazine* noted in 1837. In a piece ominously headed 'A Scourging Soliloquy about the Annuals' he dismissed *Finden's Tableaux*, edited by Mrs Hall, as light weight and unimpressive, adding that 'Mrs Hall's own efforts are light, sketchy and agreeable; but she should eschew romance, and leave all such lofty flights to *Pelham* and the Vauxhall balloon, which, consisting of such gaudy colours and air, an ascend to any elevation'.[31] (The reference is to *Pelham*, Edward Bulwer-Lytton's novel of fashionable life.) A more pointed criticism of Mrs Hall's work in the Annuals had been made by the reviewer in the *Dublin Penny Journal* in a notice of the *Amulet* for 1835;

> We think Mr Hall has no great right to be much obliged to his '*cara sposa*' for her efforts in his service, as it occurs to us that they are by no means the best of her stories that she gives him for his *Amulet*. However, 'The Drowned Fisherman' in the present volume is not one of her worst, but 'The Old Clock' should have been allowed to remain in the anti-room [sic] where she discovered it.[32]

The Annuals and Gift Books were not the only publications in which Mrs Hall's stories, tales, sketches and articles were printed.

She was lucky in that the nineteenth century was the age of the periodical – it has been estimated that thousands of journals, gazettes, magazines, weeklies and reviews, all with claims to be considered 'literary', were published in that era and her work found a home in many of them. Periodicals to which she contributed ranged from the popular *Chambers's Journal* to the more rarefied and radical *Westminster Review*. Not every editor was instantly impressed by her work, however, and John Stuart Mill complained in a letter to his friend John Robertson about an article submitted to him in 1839 by Mrs Hall when he was editing the *Westminster Review*. It was meant to appear as part of a series 'The Heads of the People', but Mill said forcefully:

As for Mrs Hall's [contribution], I have not yet dared to touch it. It is beyond all measure bad, and impossible to be made better. It has no one good point but a few of the stories towards the end, and those are told cleverly and with sprightliness, no doubt, but in the tone of a London Shopkeeper's daughter.

If I have my way we shall reject it totally, but if you could possibly suggest to me any way of making it endurable I should be happy to try them.

One thing I am determined on: nothing shall go to Paris under my sanction and responsibility showing such ignorance and such cockney notions of France and French matters as this does.[33]

The article appeared in the October issue of the *Westminster Review*, but whether or not Robertson was able to do some discreet rewriting it is impossible to say, for it is in no way remarkable or in any sense objectionable.

Fortunately for Mrs Hall, her 'sprightliness' and cleverness appealed to enough editors and readers to ensure that she was constantly in demand. Her journalistic output was prodigious, and no doubt the financial pressure caused by Mr Hall's misfortunes was a great spur to activity. The Second Series of *Sketches of Irish Character* was published in 1831, and although some of the Sketches had already appeared in the Annuals, Mrs Hall wrote several new ones for the book publication. She continued to write Irish sketches, this time for the *New Monthly Magazine*, based on visits to Ireland which she had made with her husband, and these appeared throughout 1834 and 1835. They were later collected, new material was added to them, and published in three volumes in 1838 as *Lights and Shadows of Irish Life*.[34] All this involved a great deal of hard work, but she also managed, in the years between 1832 and 1840 to write four full-length novels, *The Buccaneer* (1832), *The Outlaw* (1835), *Uncle Horace* (1837) and *Marian* (1840), all of which were well received and added to her standing as a writer. A selection of her numerous short stories, written for various periodicals appeared in 1835 under the title *Tales of a Woman's Trials* and was also a popular success. In this period she also wrote three plays, *The French Refugee, Mabel's Curse* and The *Groves of Blarney*, (the latter two being stage adaptations of her Irish stories) and edited the *Book of Royalty* as well as doing journalistic and other work. One can only marvel at the woman's energy. Mr Hall was equally busy – he once wrote a simplified history of France in 18 days for the publisher Henry Colburn, then collapsed with 'brain fever' – but did not achieve the success which his wife enjoyed. He did not attempt fiction, but wrote essays of literary and artistic criticism and some solemn poetry, which would have rated more highly among serious-minded people than mere story-telling. He may have been able to repair whatever damage Mrs Hall's success had done to his self-image by pointing out the transient nature of her popularity and the impermanence of her stories and sketches. In the event, he was right.

The Annuals went out of fashion in the closing years of the fourth decade, but the periodical press went from strength to strength, providing ever-new markets for Mrs Hall's work, and when her husband settled down to his career with the *Art Journal* she provided regular articles for that paper. For a time, too, there was a market in Irish periodicals for her tales, but it wasn't a very large one. The story of Irish periodicals in the first half of the nineteenth century has been told by Barbara Hayley in her lucid contribution to the collection *Three Hundred Years of Irish Periodicals*, published in 1987. She demonstrates that for the first thirty years of the century the only magazines that flourished were religious ones.[35] Crude anti-Catholic propaganda was countered by crude anti-Protestant polemics but new literary journals began to appear in 1830, and the first of these, the *Dublin Literary Gazette & Weekly Chronicle*, which later changed its title to the *Irish National Magazine*, printed original work by Mrs Hall. In a letter accompanying her sketch 'Kate Connor', which was published in the magazine in January, 1830, she said to the editor:

> It gives me very sincere pleasure to proffer you my first contribution to the *Dublin Literary Gazette*, and to offer you, as an Irishwoman, my very grateful thanks for having commenced an undertaking – the success or failure of which will make me either proud of or ashamed of my country.
> If Scotland *can* and *does* support *two* weekly literary journals, it would be melancholy indeed, if in Ireland *one* such publication, and that so spirited a

one as is now about to issue from the press of Dublin, did not prosper. My humble, but cheerful attempt, may, I hope, do good in one way – it may induce some of the many hundreds of our countrymen and countrywomen who are reaping gold and golden opinions on this side of the channel, to use their pens in forwarding a work which must decide the yet undetermined point – whether in Ireland there is spirit, energy, and national feeling enough, to sustain a literary periodical.

Pray command my services in any way, and believe me,

Faithfully yours,
Anna Maria Hall[36]
59 Upper Charlotte St.
Fitzroy Square,
December, 1829.

The expression of nationalistic pride and the good wishes were genuine, no doubt, but there must also have been a journalist's understandable pleasure at finding a new market for contributions. Also, publication in Irish periodicals would reinforce her standing in England as an interpreter of the manners and customs of what she called her 'native land'. Another of Mrs Hall's sketches appeared in the magazine before it ceased publication in 1831, and her work was in excellent company – fiction and poetry by William Carleton and Samuel Lover, thoughtful reviews of work by new Irish writers, scientific papers read to the Royal Irish Academy and the Royal Dublin Society, and surveys of Irish ballads and ballad singers. This expression of cultural nationalism untainted by religious bigotry was something quite new in Irish publishing, a point emphasised by the editor, Charles Lover, when he changed the title of *Dublin Literary Gazette* to the *Irish National Magazine*. The management of the magazine was Protestant, but it can fairly be said that in general the publication did not fall into the religious trap. On the Catholic side the equivalent of the *Irish National Magazine* was the *Irish Monthly Magazine of Politics and Literature*. It lasted for two years, 1832–34, and claimed that it would encourage Irish talent in literature and the arts, and be neither Whig nor Tory but simply Irish. Barbara Hayley comments, 'It did not live up to this ideal, being rather partisan; the interesting thing is that it should have claimed to do so'.[37]

The failure of these magazines and others did not deter the founders of another literary journal, the *Dublin University Magazine*, which, against the odds, ran with conspicuous success from 1833 to 1877 and gained a distinguished place not only among Irish journals but also among English and Scottish ones. It was overtly Conservative and Protestant, but in spite of its Establishment bias, it managed to project a sense of Irish national identity. Its "Irishness" was a selling point in Britain in precisely the same way as was Mrs Hall's – the "Irishness" that appealed to those English and Scottish readers who were amused, entertained and puzzled by their strange neighbours across the water. It is not surprising so, that some of Mrs Hall's stories and sketches should appear in its pages, side by side with contributions from William Carleton, Charles Lever, Isaac Butt, Caesar Otway, Charles Petrie and James Clarence Mangan. She was claimed by the magazine as one of their own – an Irish writer of outstanding talent and delicate feminine sensibility. Irish identity was being established not only in expensive journals such as the *Dublin University Magazine* and its Catholic counterpart the *Dublin Review* (published in London, but backed by a strong body of Irish talent, and lasting from 1836 to 1969) but also in the popular penny magazines which began to appear in the early 1830s. Here too, Mrs Hall found a market for her work, and sketches and stories by her were published in the *Dublin Penny Journal* (1831–37), the *Irish Penny Magazine* (1833–34) and the *Irish Penny Journal* (1840–41). All in all, these were the good years for her professionally, and her reputation in literary circles in England and Scotland was enhanced by her publication in Irish journals as diverse in price, content and intellectual stature as the *Dublin University Magazine* and the *Irish Penny Journal*.

THREE

Teaching – The Taste of the Times

Mrs Hall's contemporary success owed a great deal to the fact that she was an enthusiastic teacher. She wanted, as she so often said, to make the Irish people known, understood and loved in England, and she also wanted to improve the Irish people themselves. This didacticism is central to her work from the beginning to the end of her writing career. This overt teaching purpose is not to our taste and as Professor Jeffares has commented, 'Her didacticism was so intrusive [that] it spoils the general effect of her work for modern readers'.[1] Quite so, but Mrs Hall's books must be seen in the context of her times, and she was only one of many popular authors whose work was deliberately didactic.

Didacticism itself has an honourable literary history in its basic meaning of 'teaching', and there is a sense in which all writing is didactic, because the passing on of information or the expression of a view is, in itself, a teaching process. The writer may not consciously have decided to teach but the fact that he has spoken at all argues a subconscious wish to enlighten others. A travel book viewed in this way may be considered didactic because the author is describing his experiences among people whose surroundings, customs and way of life are unfamiliar, and he is dispelling ignorance. Then there are works which are overtly didactic in function, where the author deliberately sets out to teach people about what he thinks they should know and should think. The knowledge he passes on to them may help them in a practical way – they may learn how best to rear sheep, for instance, or may be shown a more efficient way of catching fish. On a higher plane, he may tell them what he knows of God, in the hope of helping them to worship Him in a more devoted way, or he may tell them how to behave towards one another in such a way that amity and justice may exist between them. Sometimes all this advice and sharing of knowledge is contained in the same treatise. This is the case with religious books; the Christian *Bible*, the Jewish *Talmud* and the Moslem *Koran*, where all that was known about the mysterious force called God by certain inhabitants of an area in the Middle East was gathered together over several centuries.

Medieval European literature was, in one sense, entirely didactic, for it stressed moral content in accordance with Christian doctrine, and always sought to enlighten and inform the reader. The Mystery Plays which were so popular up to the end of the sixteenth century used the dramatic form as a teaching vehicle, and in them the whole story of Creation, the Fall and the Redemption was enacted. These plays constituted a genuine theatre of the people and involved large sections of the community who were taught essential lessons in a most palatable and acceptable way. Among the more leisured and learned classes in England, didactic poetry in the Greek and Latin sense became very popular, and was used by, among others, Gower, Tusser, Giles Fletcher and Joseph Hall, and it reached its apogee in the eighteenth century with the publication of James Thomson's *Seasons* (1730). The poem combines its information about animals, plants, geology and meteorology with a high content of didacticism in the form of tales illustrating moral points. The publication of *The Seasons* coincided with the new interest in the picturesque aspects of nature and the romanticising of landscape, when the hard work of living off the land was ignored in favour of the aesthetic joy its contemplation could inspire in the hearts of those who could afford to appreciate it. This romantic view of the countryside was also expressed by William Cowper in his poem The *Task* (1785), where the natural beauty was a paradigm of primal innocence, an attitude summed up in the tag 'God made the Country, and Man made the Town'. The combination of practical information and moral lesson was also found in the poetry of George Crabbe, especially in *The Parish Register* (1807), *The Borough* (1810), *Tales of the Hall* (1822), and *Tales in Verse* (1812), but unlike Thomson and Cowper, Crabbe was no romantic. He saw rural life for what it was, an existence of grinding labour for small reward, and his subjects were the poor and obscure men and women who lived in the countryside.

As the fashion for poetry with a message was fading, the novel was rising in popularity, and it proved to be the perfect teaching vehicle. A novel was free from the conventions of poetry and was compendious enough to contain moral reflections, religious injunctions and practical information. A lesson could be clearly stated or could be implied by the fate of the characters featuring in the story, and many of the great novels of the eighteenth century had such lessons to teach. In dedicating his comic novel *Tom Jones* to George Lyttleton, Fielding declared his purpose:

To recommend goodness and innocence has been my sincere endeavour in this history ... and it is likeliest to be attained in books of this kind; for an example is an object of sight ... Besides displaying that beauty of virtue which may attract the admiration of mankind, I have attempted to engage a stronger motive to human action in her favour, by convincing men that their true interest directs them to a pursuit of her![2]

Samuel Richardson speaks in the Preface to *Pamela* of 'inculcating religion and morality in so easy and agreeable a manner as shall render them equally delightful and profitable',[3] and the sub-title of *Virtue Rewarded* is a clear indication of the high moral purpose of the book. In the Preface to *Clarissa* he states that one of the principal aims of the book was to 'caution parents against the undue exertion of their natural authority over their children in the great art of marriage: and children against preferring a man of pleasure to a man of probity, upon that dangerous but too commonly perceived notion, that a reformed rake makes the best husband'.[4] Not all authors avowed their didactic aims so openly but the teaching intent was clear, not only in such examples as Goldsmith's *The Vicar of Wakefield* (1766) and Johnson's *Rasselas*, (1759) but in the popular novel of sensibility, *The Fool of Quality* by Henry Brooke (first published in 1765–70), of which John Wesley said that 'the thinking reader is taught, without any trouble, the most essential doctrines of religion ... It perpetually aims at inspiring and increasing every right affection: at the instilling gratitude to God, and benevolence to man. And it does this not by dry, dull, tedious precepts, but by the highest examples that can be conceived'.[5]

The presence of 'tedious precepts' did not prevent Hannah More's semi-novel *Coelebs in Search of a Wife* (1809) from being a popular success, although the critic in the *Edinburgh Review* in a notice of the work in 1809 compared its teaching methods unfavourably with those of Richardson. Dismissing the novel as a 'dramatic sermon' he said: 'The book is intended to convey religious advice; and no more labour appears to have been bestowed on the story than was merely sufficient to throw it out of the dry didactic form, whereas *Sir Charles Grandison* ... teaches religion and morality to many who would not seek it in the productions ... of professional writers'.[6] Hannah More has been described by a modern commentator, Angus Ross, as 'a good representative of a lady of "culture", that is, the literary, moralistic, didactic, practical side of eighteenth-century civilization',[7] and by her teaching through fiction she fulfilled what another modern critic, B.G. MacCarthy, called the 'demand that women writers be either morally didactic, dilettante, or distressed'.[8]

One of these writers, Mrs Inchbald, was a favourite of Maria Edgeworth, who refers in a letter to Miss Ruxton to her fourth reading of Mrs Inchbald's *A Simple Story*, (1791) a novel urging the necessity of a proper education.[9] The novels of Jane Austen are not immediately didactic, but lessons on personal morality or behaviour may be drawn from them, especially from *Mansfield Park* (1814) where, in the view of Gilbert Ryle, she is demonstrating her specific moral ideas drawn from Shaftesbury.[10]

Whatever form it took, the teaching function of the novel was well established by the beginning of the nineteenth century, and fiction was expected to convey lessons from which readers could benefit, and a study of book reviews in the leading English, Scottish and Irish literary magazines for the period 1800–1852 shows that this expectation influenced the attitudes and literary judgments of the reviewers. Their main preoccupation was with the 'moral or religious tendency' of a book, and the effect it would have upon readers. Again and again the phrases recur – 'calculated to do good', 'ingenious and instructive moral lessons', 'admirable moral purpose'. In an essay on novel reading in the *Irish Farmers' Journal and Weekly Intelligencer* in 1813, the anonymous author put the prevailing view very clearly: 'The true business of the novel writer is to give faithful and interesting pictures of human life, and to inculcate through the medium of the narrative the true principles of religion and morality'.[11] Any other type of novel was dangerous to its 'devotees', especially women, who would be made 'helpless and irritable', and would suffer from irresolution.

When reading these reviews one might wonder if moral teaching was the sole criterion of a book's worth. Yet this is not the case. The novel was expected to teach, but the manner of its teaching was scrutinised carefully. Indeed, even the matter of the book under review could cause unease, as in the propriety of using fiction to convey a specifically religious or doctrinal lesson, which was adverted to by the writer in the *Eclectic Review* in 1806, who reviewed *The Morality of Fiction* by H. Murray. He concluded that it was in order to use fiction in this way, citing the example of 'the Divine author' himself, so that 'such a mode of conveying instruction, under certain restrictions, may be perfectly legitimate'.[12] These restrictions, in the opinion of the critic who reviewed Henry Lacey's *Life of David, King of Israel* in the same periodical some years later, should be applied to stories about Biblical characters, and, he added that 'there are but very few instances in which we should be disposed to adopt fiction as a fit vehicle for Divine truth'.[13] This critic showed prescience, realising that works of fiction in which figures from the Bible played a part, could dissipate the moral effect achieved by the original narrative and could place it in the ranks of mere novels, rather than in its pre-eminent place as the book in which all wisdom was to be found. There was, of course, no objection to a novel that combined religious truths and 'practical wisdom' as did the Reverend Croly's *Tales of the Great St Bernard*, reviewed in the *Gentleman's Magazine* in 1828, and these were the constituents that particularly recommended it to the reviewer, who foresaw that its publication would enforce 'sound moral precepts'.[14]

Religious novels *per se*, however, were not certain of a good reception. Speaking of Mrs Sherwood's novel *Roscobel* in the *Athenaeum* in 1831, the reviewer complained about the poor quality of the writing in some religious novels: 'The tales are bundles of incidents bound together by statements of religious sentiment. Even religion itself is seldom treated with adequate dignity ... instruction is rarely interwoven with the fabric of the fiction itself, but appended as a fringe'.[15] Conversely, the lack of a religious purpose was, in the view of some critics, a great drawback. Although most reviewers applauded Miss Edgeworth's 'high moral purpose', her lack of conventional religious teaching was noted. The *Monthly Magazine* in a very strong attack on her works in 1826, not only found her style 'cold, laborious and unnatural', her mind 'mechanical', and her wisdom 'the dry and crippled manufacture of old-maidism and governess-ship', but criticised her work as being 'totally founded on selfishness, under the name of prudence', and believed that she had 'dispensed with the influence of religion'.[16] The writer who reviewed Miss Edgeworth's works in the Irish Protestant *Dublin Monthly Magazine* in 1830 agreed with this assessment, deprecating her 'Utilitarian theory of virtue', and accusing her of being 'blameably silent on the subject of religion',[17] while the mouthpiece of more violently anti-Papish Irish Protestantism, the *Christian Examiner*, in its review of *Irish Priests and English Landlords* by the author of *Hyacinth O'Gara* (George Brittaine) in 1830, also referred to Miss Edgeworth's failure to promote religious principles in her writings. 'Her melancholy indifference to religion ...', said the reviewer, 'precluded the shadow of a hope that anything would ever be effected toward the spiritual amelioration of her country's interests', and it was fortunate that in recent years some writers, including the author of the novel being reviewed had shown 'intellectual ability plus real piety.'[18]

While there was no disagreement among literary critics about the worth of the lessons taught by fiction there was some confusion about the application of the word 'didactic'. Was it *ipso facto* a term of approbation or should it be used in a derogatory sense? In an article in *Blackwood's* in 1819 on novel writing in general there is an attempt to clarify the issue:

> The manners and morals of the middle-classes have ... been handled in works, which are not written like the highest novels, for the sake of recording the developments exhibited by the human mind, but what might be called "moral" novels, because they have generally a didactic purpose, relating to existing circumstances and are meant to show the causes of success or failure in life, or the ways in which happiness or misery is produced by the management of the passions and affections.[19]

This would indicate that, in this critic's view, didactic novels are of a lower order than others in an artistic sense, but then he goes on to speak of novels of 'passion and sentiment' for example, *Werther* and *The Nouvelle Heloïse*, stating that they are not "didactic" works; for 'no person reading them, ever picked up rules of practical prudence, or gained more control over his passions', thus implying that didactic novels have a useful effect. *Sir Andrew Wyllie* by the 'Reverend Michah Balwhiddie' (John Galt), was praised for its teaching in a notice in the *Edinburgh Review* in 1823 in a way which leaves no doubt that the reviewer considered 'didactic' to be a term of praise: 'There is ... a more distinct moral, or unity of didactic purpose, in most of his writings than it would be easy to discover in the playful, capricious, and fanciful sketches of his great master',[20] the reference being, of course, to Sir Walter Scott. As late as 1845 the term was being used in its laudatory sense, thus the novels of Heinrich Zschokke, the German writer, who enjoyed a brief popularity in England, were praised in *Tait's Edinburgh Magazine* with the comment that 'the most characteristic, and indeed, the most interesting of Zschokke's novels are those which are avowedly didactic ... His best stories are narrative sermons, of which a social truth, or a proposition represented as the truth, is the text'.[21] Moral and didactic go hand in hand, as in the review of *Florestan* in the *New Monthly Magazine* in 1839, which noted that the story had a 'purely didactic and moral object, and that object is an excellent one – the making known the true condition of the English peasantry, and educing from that condition the means of their amelioration'.[22]

The term 'moral' was, in spite of the forgoing example, generally synonymous with 'didactic' in the language of the literary critics and a typical use of the term and an insight into the thinking of the critics of the period is found in the *New Monthly Magazine* of September, 1837. In the regular feature, 'The Conversazione or the

Literature of the Month', the Rector, speaking of *Stokehill Place, or the Man of Business*, by Mrs Gore, says:

> It has been a doubt whether a novelist should have a moral in view, and unquestionably if the moral is made the groundwork of the story, the story will be dull. No man reads a novel for the express purpose of being made wiser; no man reads it as a guide to his conduct in life, or as the controller of his passions; and yet the novel is imperfect which does not contain all of these results; which does not impress some great principle or moral without the ostentation of wisdom, increase the force of our thought without affecting to sharpen the keenness of our understanding; and, in the simplicity of its incident, add to our knowledge of the motives that form the mainspring of human life.[23]

Didactic purpose, therefore, included that of moral regeneration, either of individuals or social groups or both. It also connoted instruction in how to behave in accordance with the laws of God and those of man. There was no question of overthrowing the existing social order, though changes should be made within it, and Maria Edgeworth, in spite of her alleged indifference to religion, was praised in the *Gentleman's Magazine* for her perceived intention 'to reform the morals of society and to call attention to ... the education of youth',[24] and she was the criterion by which other writers were judged. Her *Tales*, in the opinion of the critic in the *Eclectic Review* who was reviewing a novel in 1815 called *Display* were 'the most ingenious and instructive moral lessons that ever proceeded from the pen of an individual',[25] and a critic in the *Edinburgh Review* praised 'that series of moral fictions ... in which ... she has combined more solid instruction with more universal entertainment, and given more practical lessons of wisdom, with less tediousness and less pretension, than any other writer'.[26] The popular novel *Tremaine* (praised in *Blackwood's Magazine* for its 'admirable moral purpose'[27] was seen by the *New Monthly Magazine* in 1825 as resembling Miss Edgeworth's novels 'in [its] excellent moral lessons touching the conduct of real everyday life',[28] and the *Edinburgh Literary Gazette* stated that *The Recluse*, 'a work which had as its professed design to paint vice in all the deformity of its native colours, while it encourages the reader to an adherence to virtue, by pointing out the native charms of morality and truth', was following 'in the tracks so successfully pursued by Miss Edgeworth, Mrs Opie and a few others'.[29] In a review of Harriet Martineau's works in the *Quarterly Review* in 1833 the author was advised to study Miss Edgeworth as a social commentator, particularly in her novels *Castle Rackrent* and *The Absentee*, the critic pointing out that there is not indeed one tale of Miss Edgeworth's but conveys some useful lesson on questions which naturally concern the economy of society. But the difference between the two writers is that the moral of Miss Edgeworth's tales is naturally suggested to the reader by the course of events of which he peruses the narrative; that of Miss Martineau is embodied in elaborate dialogues and unnatural incidents with which her stories are interlarded and interrupted.[30]

The artistry of Miss Edgeworth was recalled in 1843 in *Fraser's Magazine* in a review of the novels of Frederika Bremer, a writer whose fiction was highly regarded at the time, when the critic commented: 'as teachers of truth through the medium of fable she [Miss Bremer] and Miss Edgeworth may be classed together, but the immeasurable superiority of the latter as a novelist places a wide gulf between them'.[31] The teaching purpose of Miss Edgeworth's works remained as an object of praise as late as 1848, when in *Bentley's Miscellany* in a memoir of the author Sir James Mackintosh, the historian and philosopher is quoted as saying:

> I thought that Miss Edgeworth had first made fiction useful; but every fiction since Homer has taught friendship, patriotism, generosity and contempt of death. These are the highest virtues, and the fictions which taught them were, therefore of the highest, though not unmixed utility. Miss Edgeworth inculcates prudence, and the many virtues of that family. Are these excellent virtues higher and more useful than those of fortitude of benevolence? Certainly not. Where then is Miss Edgeworth's merit? Her merit, her extraordinary merit both as a moralist and as a woman of genius – consists in her having selected a class of virtues far more difficult to treat as a subject of fiction than others.[32]

Concurrent with this assessment of Maria Edgeworth's work, however, was a resistance to her method of achieving her purpose. A distinction was made between matter and manner and while acknowledging that her aim was laudable, some of the critics expressed their dislike of her method, which was 'occasionally too didactic, too wise', according to the *Athenaeum*[33] and the *Edinburgh Review*, overall an admirer of her work, though noting that her aim was to 'correct fatal errors of opinion ... to display wisdom and goodness at once in their most engaging and familiar aspects and to raise humbler virtues to their proper place', declared that

> it is to the unrelaxed intensity of this pursuit that we think almost all her faults are to be referred. It is this which has given to her composition something of too didactic a manner and brought the moral of her stories too

obtrusively forward. This defeats its own ends because adults do not want instruction in such a direct and officious form.[34]

The *Quarterly Review* in its notice in 1821 of *Northanger Abbey* and *Persuasion* stated that 'the immediate and peculiar object of the novelist, as of the poet, is to please', and criticised works which made direct attempts at moral teaching, while allowing that a 'certain portion of moral instruction must accompany every well-intentioned narrative'. Jane Austen, in this respect, was above criticism, and her skilful construction was praised as 'her moral lessons spring from the circumstances of the story'. This was not the case with Miss Edgeworth who was 'too avowedly didactic'. It seemed as if she had first thought of the moral and then made a story to fit around it. This lessened her value as a teacher. She 'would instruct better if she kept the design of teaching more out of sight, and did not so glaringly press every circumstance of her story, principal or subordinate, into the service of a principle to be inculcated, or information to be given'.[35] 'Didactic' is being used in a pejorative sense in both of the reviews quoted, and it was used in this sense by some reviewers throughout the century, though sometimes confusingly defined. An early use of the term in an unflattering way was in a review of a book called *Precaution*. A writer in the *New Monthly Magazine* in 1821 noted that one of the characters in the novel was described as labouring 'under the disadvantage of a didactic manner', and the critic felt that the same could be said about the author.[36] Yet, in the same issue, the author of a poem called *Echoism* was praised for being 'didactic without being tedious', so underlining the critical confusions.

A desire to teach might, as in Maria Edgeworth's case, spoil the story, as the reviewer of Lady Morgan's *The O'Briens and the O'Flahertys* pointed out in the *Monthly Review* in 1827: 'She condescends to employ the machinery of a romance only as a convenient vehicle for announcing a great "moral" lesson. Wholly engrossed by her didactic purpose, she overlooks every other consideration.'[37] The 'mischievious and unfeminine doctrines' of Harriet Martineau were expressed in 'dull, didactic dialogues' in her *Illustrations of Political Economy*, according to the review in the *Quarterly Review* in 1833,[38] and there the linking of 'dull' and 'didactic' emphasises the derogatory use of the latter term. Her novel *Deerbrook* was seen in *Blackwood's* in 1840 as more 'feminine' but here, too, her teaching method came under attack, in an explicit condemnation of didacticism. The novel was considered greatly superior to her earlier *Tales* because it was not so overtly moralistic and concerned to teach, but made its points in a more subtle way. 'Didactic poetry', said the reviewer, 'is no poetry except when it forgets to teach. The *Georgics*, of which the true subject is the praise of country life, would form a perfect poem if it were possible to remove from them the agricultural precepts with which they are encumbered.'[39]

A new use of the word 'didactic' was made by the writer who reviewed *Rose, Blanche and Violet* by G.H. Lewes in *Fraser's Magazine* in 1848. He posed it in opposition to realistic, saying that 'Mr Lewes very wisely abandoned the idea with which he commenced of writing a didactic fiction – he wrote instead about life as it is'. In the same review the critic stated that 'for absolute teaching we should not look to the novel. To amuse and moralize, to weave a tale and develop a theory, to paint the passions, and, at the same time, to lay the foundations of a new philosophy are undertakings altogether incompatible', but he insisted that the novel should become the 'fearless though informal censor of the age and hold society in severe check by mercilessly exposing the errors, weaknesses, absurdities, excesses and even crimes which disfigure and disturb it'.[40] This is a recognition of the direction which the teaching novel has taken and an implicit commendation of the works of Dickens, Disraeli, Charles Kingsley and Elizabeth Gaskell, although the reviewer of *Mary Barton* in the *Westminster Review* in 1849 did not foresee immediate practical results from the teachings of these authors, saying of Mrs Gaskell's novel that 'it embodies the dominant feeling of our times – a feeling that the ignorance, destitution and vice which pervade and corrupt society, must be got rid of. The ability to point out how they are to be got rid of, is not the characteristic of this age. That will be the characteristic of the age which is coming'.[41] This observation on the ability of authors to identify problems while being unable to put forward suggestions for their solution is one which had been made earlier in the *Dublin University Magazine* in a review of Mrs Hall's novel, *The Whiteboy*, in 1845, and it was equally pertinent in that case. The recognition that a didactic effect is a worthwhile object is taken for granted in a survey, 'The English Contemporary Literature of England' in the *Westminster Review* in 1852, but the manner in which it is achieved is now under attack. Dismissing the novel *The Fair Carew*, the critic says, 'It is of that class of religious novels which aim at a didactic effect by an inflated style of reflection; and by melodramatic incident, instead of faithfully depicting life and leaving it to state its own lesson as the stars do theirs'.[42]

A similar critical opinion was expressed in *Fraser's Magazine* in

1850 in a general article on fiction. The critic put forward the view that novels were for amusement, not for information or instruction. Referring specifically to Harriet Martineau's *Deerbrook*, he reminded readers that 'a few years back it was quite the fashion to set all kinds of truths in a filigree work of fiction'. Having denounced 'all those who, upon pretence of entertaining, design to convince us to their politics, or convert us to their creed', and novelists who falsify historical truth in their fictions, he goes on to enter a protest against "moral tales", that is to say: 'tales whose professed end is the inculcation of a moral lesson; such as *Patronage* and *Manoeuvring*, where the same indisputable truth comes up and confronts you at every turn of the narrative. Life teaches its lessons in an obscure and complex way; we find out its truths for ourselves, and this is the way in which the novelist should teach us'. He is, in fact, accepting the novelist as teacher, and what he is objecting to is the way in which this rôle is played, and to the ostentation of the teaching purpose. As far as he was concerned, Aesop's *Fables* were the only acceptable moral tales and any modern attempt to use the same formula in fiction was doomed. He continued: 'Miss Edgeworth's "instructive" works are fast descending into oblivion and the pastrycook, and Miss Austen's tales, heavily weighted by their titles, are kept afloat only by the genial and natural life within, which belies the frigid exterior'. He concluded by praising a now-forgotten novel, *Julia Howard*, and commending its author, a Mrs Martin Bell, because she had made 'no attempt at the didactic'.[43] It was in *Fraser's Magazine* also that a critic, some years later, in 1852, reviewing *Reuben Medlicott* by M.W. Savage, insisted that it was a mistake 'to suppose that an attractive story is not essential to the success of a novelist even of the didactic order. On the contrary, it is an indispensable condition'.[44]

In a review of English novels in *Fraser's* in 1851 the critic, commenting on the immorality of French novels, pointed out that they were never dull, and added, 'We cannot help regretting that ... our English novelists, who, for the most part, write unexceptionable morality, should not be able to render it a little more amusing. It is a pity that morality should be rendered so excessively stupid on this side of the Channel'. He goes on to make an attack on the lack of artistry in English novels, saying unequivocally: 'We are too didactic. Thinking too much of the moral, and too little of the story, through which it is enforced, we suffer the end to overwhelm the means', and adds as criticism of the novel *The Tutor's Ward*, that 'the excess in which the didactic element prevails over the dramatic is a great fault'.[45] This criticism of the didactic aim, which accepts its matter but not its manner is mild compared to that levelled by a writer in *Blackwood's* at the whole class of what he calls 'philosophical novels'. These are 'works where some theory or dogma is expressly taught, where a vein of scholastic, or political, or ethical matter alternates with a vein of narrative and fictitious matter'. He dislikes them all and cannot see how they could be successful in any sense: 'Either one is interested in your story, and then your philosophy is a bore; or one is not interested in it, and then your philosophy can gain no currency by being tacked on'. Even if the 'narrative and didactic portions of such a book are equally good, it is still essentially two books in one, and should be read once for the story and once without'. Fortunately, he added, it was easy to do this with some novels because of the technique used. In *Tremaine*, he recalled, 'the didactic portion has sunk like a sort of sediment, and being collected into a dense mass in the third volume, could easily be avoided'.[46] This is a very different judgment from that made by a reviewer in *Blackwood's* twenty-five years earlier who thoroughly approved of *Tremaine* because of its 'admirable moral purpose and tendency throughout', without differentiating between the story and the lesson.

In 1845, in a general article about popular literature, the *Eclectic Review* frowned on the 'mischievous excitement of indiscriminate novel-reading'. There were very few novelists whose works were worth reading or buying – among them were Sir Walter Scott, Maria Edgeworth, Jane Austen, Frederika Bremer and Charles Leaver [sic] – and the circulating libraries, those 'gin palaces of the reading world' were blamed for the debasement of public taste. The writer pleaded for a new type of literature, stating that

> after the twelve hours of strenuous attention to business in the shop, the warehouse or the office, the mind is not always able to seize on grave or solid matters of importance. It requires to be soothed, to be entertained and refreshed by light, joyous and vivid imagery. Under these circumstances, fictions that describe actual life, or that give actual life and character new hues, and draw from them without any didactic assumption, new lessons, become not only our entertainers, but our friends and instructors.[47]

The debasement of public taste had been noted earlier and those who catered to it had been severely criticised in the *Athenaeum* in 1838. Lamenting the depiction of vice and squalor in Harrison Ainsworth's novel *Jack Sheppard*, (1839) the reviewer contrasted his work unfavourably with that of Dickens:

> If Boz has depicted scenes of hardened vice, and displayed the peculiar

powers of degradation which poverty impresses on the human character under the combination of a defective civilization, he is guided in his career by a high moral object; and in tracing what is loathsome and repulsive he contrives to enlist the best feelings of our nature in his cause, and to engage his readers in consideration of what lies below the surface.[48]

The lowering of public taste, in this critic's view, resulted from several causes, including an estrangement from nature, and the consequent viewing of 'all individualities in one common character, cold, monotonous, superficial and polished ... but hard and hollow'. The crowding together of people in towns, the difficulties of the struggle for existence and the spread of a 'delirious spirit of devotion' (a reference to Methodism rather than Tractarianism, given the different social appeal of each movement), also contributed to the coarsening of popular taste.

Considering the continued uneasiness about the bad effect novel reading could have on the character, it is not surprising that the Evangelical periodical, the *North British Review* should have found it necessary to explain to its readers what its policy was in regard to fiction. In the Preface to the first issue in 1844 it was announced that when taking notice of novels 'the moral and religious tendency of works under review, will always be the first subject of consideration'.[49] In view of this policy it is clear why novels by Mrs Maitland, *The Ogilvies*, and *Olive*, were welcomed for their high moral tone, although the reviewer had little else to say about them. He pointed out that 'one among many modes, at least, if not the highest or most direct, of inculcating truth and encouraging goodness, is a good novel'. In the same review was a notice of *Mary Barton*, and although the critic did not care for the novel, and thought that Mrs Gaskell took too sympathetic a view of the improvident poor of Manchester, he approved of her method of teaching and felt that most readers would be happy with the book's indirect, unconscious mode of teaching through the medium of facts, in preference 'to long-winded interruptions to the plot in the shape of didactic dialogues'.[50] The value of teaching through fiction was self-evident, according to a writer in the *Prospective Review* in 1851. In a review of *Yeast* by Charles Kingsley under the general heading of 'Polemical Fiction', he accepts that the majority wants to be entertained, and for them 'the pill of truth must be gilded or it will be rejected', but he believes that

there are many who take up books for higher purposes than amusement ... and there is enough, and more than enough, among the professed allegories of older date, and the didactic fiction of the late classical age of German literature and the cognate productions of French and English genius to sustain a thesis in favour of prose poetry with distinct moral and metaphysical aims. The character of the most successful and elegant modern prose fiction sustains us in the belief that by devotion to the highest ends, this class of literature gains rather than loses in artistic effect.[51]

Imagine, he says, Bunyan without Calvinism or Schiller or Goethe without transcendentalism. He ends with a plea for 'realism', believing that the 'most finished and effective form of didactic literature is truthful fiction', a paradoxical phrase which had already been used by William Carleton in 1845 in the Preface to his *Art Maguire*, a cautionary tale for peasants about the evils of intemperance.[52]

An attack upon the whole concept of didactic fiction was made in 1845 in the *Dublin University Magazine*, the Irish Protestant and Tory periodical, by Sir Samuel Ferguson in his review of *The Whiteboy* by Mrs S.C. Hall, and *Tales for the Irish People* by William Carleton. He disapproved of the genus even when it was skilfully done. There was an element of deceit in it and it was 'a base thing to entrap a reader by an agreeably-written love scene or exciting adventure, into the admission of a dogma which, if fairly stated side by side with the adverse proposition, would depend for its admission or rejection, on its being found agreeable or otherwise, not to the imagination but to the judgment'.[53] He is, of course, admitting the power the novel has to influence people's thinking, and is expressing a genuine worry about how this power may be exercised. Although, in this particular instance, the reviewer is speaking of books with a political message the point he is making is an ethical one, and he is calling for the novel to be placed firmly in its artistic position where it may address 'the heart and the fancy', without any fear of its emotional appeal clouding the reason of its readers. The impression gained from reading this review, and the others with their conflicting views of moral instruction in the novel, is that while there was a convention that such instruction should exist and that authors and readers alike followed the convention, the professional critics were in something of a quandary. They could not object to sound moral teaching, either in their own *personae* or as visible literary figures but they were uneasy about the artistic damage that such teaching might do. It was not difficult to condemn novels that were so clumsily written that the teaching obtruded on the narrative at every turn but when a book was so cleverly crafted that its didactic effect was felt only after it had been read, enjoyed and assimilated, there was a real problem of critical evaluation, as well as the ethical one identified in the *Dublin University Magazine*.

There was no such difficulty in assessing the practical form of didacticism, and early novels that conveyed information about people and places were looked on kindly by the critics. It was a merit in a novel to give such information even if it gave but little else to the reader. For example, a novel called *Malpas* was noticed in the *Gentleman's Magazine* in 1822 as giving the reader 'a copious display of ancient manners',[54] and the *New Monthly Magazine* in 1826 commended *The Prophetess*, a work about sixteenth-century Italy, as worth having because it gave as much information as a guide book,[55] and the *Memoirs of Captain Rock* was welcomed by the *Westminster Review* in 1824 because 'through it a knowledge of the evils of Ireland may be gained easily'.[56] In the following year a reviewer in the same periodical greeted the publication of an attack on abolitionism in the form of a novel, *Hamel the Obeah Man* by agreeing with the main thesis (slaves do not wish to be free, and even if they were they would not know what to do with freedom), and adding that extra entertainment could be gained from the fact that the scenes and manners described were new[57]. This appetite for practical information was considered no mere curiosity by the literary critic of the *New Monthly Magazine* who recommended Lady Morgan's *The O'Briens and the O'Flahertys* (1827) as 'valuable to those who read for higher purposes, as containing a succession of rich and various pictures of Irish characters, feelings and manners'.[58] 'Higher purposes' therefore, could include the amassing of information, and the works of Maria Edgeworth, Lady Morgan, Gerald Griffin, the Banim brothers, Carleton and Mrs Hall, were all in their turn praised for the descriptions of Irish life and the delineation of Irish character which each of them gave.

The debate about didacticism, the opposing views and the confusion all show how seriously the subject was taken during Mrs Hall's lifetime. It is significant that nowhere (with the exception of the Ferguson review in the *Dublin University Magazine*) is she rebuked for emphasising her teaching purpose. Even Ferguson does not criticise her method or her style, it is her *aim*, and that of all those who write novels with a purpose, that arouses his indignation. We must conclude so, that Mrs Hall's didacticism was not considered intrusive by her contemporaries, and that her teaching was not only acceptable (in an English context, whatever about an Irish one) but was seen as being entertainingly done, spiced with humour, heightened with drama, touched with pathos, and unshakably firm and unequivocally right.

FOUR

Sketches of Irish Life – The Voice of the Colonist

When Mrs Hall's first two collections of Irish sketches appeared in 1829 and in 1831, the English perception of the Irish peasant was of a figure barely human. He was dirty and ragged, lived in a hovel which he shared with his pig, ate potatoes and drank whiskey. He was lazy and superstitious, got drunk, loved to fight, and was dangerous when aroused. His speech was in an English which was so distorted as to be comic, but he sometimes displayed a low cunning. He did have some good points, but they were more animal than human – fidelity, affection and an amazing response to kindness. It wasn't even necessary to go to Ireland to meet some of these people – drunken Irish labourers, their sluttish wives and their hordes of children swarmed in every English slum.

Mrs Hall's portrayals of the Irish peasant in many ways conformed to this image. The physical appearance of some of her characters was in the tradition of figures in low-life Dutch paintings by such artists as Ostade and Jan Steen and anticipated the later cartoons in *Punch*. Larry Moore, for instance, the incurably lazy and feckless boatman, is described thus:

His hat is a natural curiosity, composed of sun-burned straw, banded by a mis-shapen sea-ribbon, and garnished by "delisk" red and green, his "cutty-pipe" stuck through a slit in the brim, which bends it directly over the left eye and keeps it "quite handy without any trouble". His bushy, reddish hair persists in obstinately pushing its way out of every hole in his extraordinary hat, or clusters strangely over his Herculean shoulders, and a low furrowed brow, very unpromising to the eye of a phrenologist: – in truth Larry has somewhat of a dogged expression of countenance, which is relieved, at times, by the humorous twinkle of his little grey eyes.[1]

Jack the Shrimp, the wanderer on the sea-shore, was a 'wild, desolate-looking creature; black lank hair fell over his face and shoulders'.[2] Molly, Father Mike's housekeeper, had 'a lean figure and scraggy neck supporting a face 'broad as a Munster potato' while her wide mouth and long, sharp teeth betokened her passion for talking and hearing'.[3] Peggy the Fisher, the wandering saleswoman in 'Lilly O'Brien' wore a 'green spotted kerchief tied over her cap – then a sun-burnt, smoke-dried, flatted straw hat – and the basket of fish resting "on a wisp of hay"'.[4] Kelly the Piper had 'a visage thin, yellow, and ghastly – except a long, pointed crimson nose, with a peculiar twist at the end, which assumed a richer colouring, shading to the very tip in deep and glowing purple', while his children 'moved about in a miscellaneous mass of brown-red flesh, white with bushy elf locks'.[5] It is notable, however, that these unflattering descriptions apply only to the lowest stratum of Irish society.

Class-consciousness as well as racism was a large factor in Mrs Hall's thinking and people higher up the Irish social ladder are painted differently. Corry Howlan, the young farmer in 'Peter the Prophet', was six feet in height, with 'an air of easy confidence, and every limb well-proportioned, face oval; teeth white and even, nose undefined as to aquiline, Grecian, snub or Roman, but, nevertheless, highly respectable; eyes large, *bien fonçee*, and expressive; brow overshaded with rich curling brown hair'[6] and his dress was respectable though highly colourful. Kathleen Ryley, daughter of a 'respectable farmer' and witness to Mabel O'Neill's curse, had a 'fine round figure, and her 'white muslin kerchief was always delicately clean, neatly meshed and carefully pinned across her bosom',[7] while the priest's nephew, Morris Donovan, hero of 'The Wise Thought' was 'the beau – the Magnus Apollo, of the parish; – a fine, noble-looking fellow'.[8] Mrs Hall's 'cottage-girl portraits' as the *Literary Gazette* termed them[9], demonstrate very clearly however, that, in her eyes, peasant beauty is of a different order to high-born beauty. Lilly O'Brien, her figure 'slight and bending as a willow wand' her skin 'transparently bright', and her hair 'a pale, shining and silky auburn', was nonetheless described by her peers as being 'almost like a born jantlewoman'.[10] Anty McQueen in 'The Bannow Postman' was 'a merry, laughing, blue-eyed lass, somewhat short, and without one good feature in her face; yet the gipsy was esteemed pretty'.[11] Contrast this with the description in 'Hospitality' of Gertrude Raymond, 'born of a noble but decayed family', whose features

when tranquil, had an expression of hauteur; her brow was lofty and expanded; her eyes deep and well set; her skin, nearly olive; her hair rivalled the raven's wing; her cheek was in general, colourless except when her feelings were excited, and then the rich blood glowed through the dark surface with the deep colouring of the damask rose, the eyes brightened, and [she] ... burst upon you in all the magnificence of beauty.[12]

A different vocabulary is used for each girl, as in the descriptions of Kathleen Ryley and Miss Caroline Johnson in 'Mabel O'Neill's Curse'. Kathleen was a 'frank-hearted, merry girl, with laughing blue eyes and a joyous countenance' while Miss Caroline's eyes were a 'clear lustrous blue' and 'her long pencilled lashes rested on the soft roundness of her delicate cheek'. She was 'slight and graceful ... her form was the perfection of symmetry yet shaped in so fair a mould that, in their youthful days, Kathleen used to boast that she could carry Miss Caroline a mile in one hand, and never know she was there', while Kate's 'round red arms' and 'sun-burnt skin' were in sharp contrast to Miss Caroline's 'polished shoulders'.[13]

The settings for more respectable members of Irish society are very different from those of their inferiors and are described in loving detail. Father Mike's kitchen, for instance, is a cheerful, cosy, well-furnished place:

Two large dogs, a cat and a half-grown kitten shared also the wide hearthstone, and enjoyed the bright cheerful light of a turf and wood fire. On an old-fashioned table, partially covered with a half-bleached cloth, was spread the priest's supper; a large round of salted beef, a silver pint mug ... an unbroken cake of griddle bread with a "pat" of fresh butter on a wooden platter, and two old bottles, containing something much stronger than water. An antique armchair, with an embroidered but much soiled cushion, was placed opposite the massive silver handled knife and fork: – all awaiting his reverence's coming. From the rafters ... hung various portions of dried meat, fish and pigs' heads ... . The dresser, which as usual in Irish kitchens, extended the whole length of the room, made a display of rich china, yellow delf, wooden noggins, dim brass, and old, but chased-silver candlesticks.[14]

Mrs Clary's kitchen in 'The Wise Thought' was 'lofty', and 'on a small round table, a cloth was spread, and some delf plates awaited the more delicate repast which the farmer's wife was herself preparing'.[15] Mrs Cassidy's parlour in the sketch 'Lilly O'Brien' was pictured in detail:

The floor strewed with the ocean's own sparkling sand; pictures of, at all events, half the head saints of the calendar, in black frames, and bright green, scarlet and orange draperies; a corner cupboard, displaying glass and china for use and show, the broken parts carefully turned to the wall, the inside of the chimney lined with square tiles of blue earthenware, and over it an ivory crucifix, and a small white chalice, full of holy water; six highbacked chairs, like those called "education" of modern days; a well-polished round oak table, and a looking glass of antique form completed the furniture.[16]

Significantly, Mrs Cassidy was no ordinary peasant – her husband had been master and owner of 'a small trading vessel', and had left his widow in comfortable circumstances.

In contrast to these simple but attractive dwellings are the homes of the poor and the improvident. Kelly the Piper and his wife Judy (who was 'one of the worst specimens of an Irish woman' ever seen by Mrs Hall) shared their dirty cabin with the pig and all paddled 'through the extraordinary black mud, which formed a standing pool round the stately dunghill that graced the door'.[17] The pig was also a resident in the McQueen family cabin and shared with the youngest child, 'a fat rosy lump' who was hardly able to crawl, every potato it took into its mouth. This pairing of the peasant with the pig was an important part of the outsider's perception of the Irish and the association of the two resulted in a transference of the porcine qualities (dirt, laziness and an innate brutishness) to the person who owned it. Significantly, the later, explicitly crude caricatures of 'Paddy' almost always showed him with his pig. (See L. Perry Curtis, Jnr's study of representations of Irish peasants *Apes and Angels, the Irishman in Victorian Caricature*, England 1971.) The candlestick in the McQueen household was a scooped-out potato, as it was in the hovel where Luke O'Brian found a dying woman. This shelter could 'hardly be called a human dwelling. It was, literally speaking. a large excavation in the earth; two boards nailed together, closed the aperture through which the wretched inhabitants entered, and a hole in the clayed roof served the double purpose of chimney and window'.[18] The farmhouse belonging to Philip Garraty, on the other hand, was 'a long rambling abode, much larger than the generality of those that fall to the lot of small Irish farmers', but it was in a sorry state. The thatch was mostly decayed, the door was hanging by one hinge, the window frames were loose and broken panes were replaced with 'the crown of a hat, or a "lock of straw"'.[19] The farm buildings and the yard were also in a state of disrepair, and pigs, hens, turkeys and geese roamed at will. None of this neglect was the result of poverty but was a consequence of the farmer's habit of procrastination.

This story, 'We'll See About It', illustrates Mrs Hall's observation that the Irish display 'a strange union of impetuosity and procrastination'. She goes on, 'They are sufficiently prompt and energetic where their feelings are concerned, but, in matters of business, they almost invariably prefer *seeing about*, to Doing'. Philip Garraty, with his wife and children, is 'employed from morning to night in *seeing about* everything and consequently in *doing* nothing'. He himself, with his 'broad, lazy-looking shoulders, and a smile perpetually lurking about his mouth, or in his bright hazel eyes [was] the picture of indolence and kindly feeling'.[20] The story is told in dialogue form, with the author questioning Philip closely about his failure to run his farm properly, while he gives evasive answers and ridiculous excuses. It transpires that, through his laziness, Philip is threatened with eviction and his once-fine farm has fallen into rack and ruin. The narrator urges him to take positive steps to find the money to save himself, but the ending is inconclusive, Garraty still promising *'to see about it'*.

This indolence or laziness of the Irish peasant class was a vital part of the English perception of the Irish peasant. Ned Lebow, in his essay, 'British Images of Poverty in Pre-Famine Ireland', sees this English attitude as a colonial one and cites Albert Memmi's *The Colonizer and the Colonized* in which Memmi notes that the vice of laziness is usually applied by the colonizer to the colonized because

nothing could better justify the colonizer's privileged position than his industry, and nothing could better justify the colonized's destitution than his indolence. The mythical portrait of the colonized therefore includes an unbelievable laziness, and that of the colonizer, a virtuous taste for action. At the same time the colonizer suggests that employing the colonized is not very profitable, thereby authorizing his unreasonable wages.[21]

This is an attractive thesis and examples in colonial writing about Ireland would seem to bear it out. Certainly Mrs Hall's writings are infused throughout with her colonial attitudes. It would have been surprising if it had been otherwise, given the circumstances of her background and upbringing. As a child she would have been aware of the difference between her way of life and that of the majority of the people in her neighbourhood. She would have been conscious of belonging to a group that was set apart by money, education and religion. Permanent residence away from Ireland sharpened her attitudes and gave an edge to her memories. Even in these early stories she used the didactic voice which developed into full cry in her later work. It was the voice of the colonist, secure in her superiority, who was entitled, morally as well as practically, to lecture the natives. In the sketch 'Larry Moore', the boatman is held up as an example of Irish laziness. He is found by the narrator 'stretched in the sunlight on the firm sand, like a man-porpoise – sometimes on his back – then slowly turning on his side but his most usual attitude is a sort of reclining position against that flat grey stone'.[22] A dialogue develops between him and the narrator, prefaced by her remark, in her authorial voice, that 'we may not moralize'. What follows, however, is explicit moralizing, the narrator haranguing the boatman and trying to make him understand how his laziness, mental and physical, is holding him back from the comforts of life and condemning him, his wife, and family, to poverty and continued insecurity. Her speech is stilted, while his is easy and fluent, and he is unmoved by her arguments, even as she traces all his problems to his laziness and refusal to plan for the morrow. Even when a prospective customer signals that he wishes to cross over in the ferry, Larry refuses to hurry, saying: '"Oh, ye're in a hurry, are ye? – Well, ye must wait till yer hurry is over. I'm not going to hurry myself wid sixpence in my pocket for priest or minister"'. A third sketch which also takes the form of a dialogue between the author and an Irish peasant is an illustration of another regrettable trait peculiar to the Irish peasantry – the habit of relying on the efforts of others rather than one's own. It is called 'Independence' and opens with a direct authorial comment; 'Of all others, "Independence" is the word that Irish – men, women and children – least understand'. Mrs Hall then goes on to question a cottager, Shane Turlough, "as dacent a boy ... as any in the world", about his misfortunes. Shane is seen 'leaning over the half-door of his cottage, kicking a hole in the wall with his brogue, and picking up all the large gravel stones within his reach'.[23] He tells the narrator of the sad consequences of his dependence on others. Because of this, his hay has not yet been cut and will be spoiled, his wife has no wheel with which to spin the flax, he has lost the lease of a new cottage, and, worst of all, his child has contracted smallpox.

Two further examples of this Irish lack of independence are given in this sketch, but they concern different strata of society – middle-class and minor gentry – and the narrative method used is different. Each tale is told in a straightforward fashion and there is no dialogue with the offenders, unlike the sketches involving peasants. The author condemns from a remove, and not face to face, the implication being that a peasant must expect direct rebuke from his superiors while those higher in the social scale may only be rebuked obliquely, and by their peers. The peasant is spoken to, they are spoken about. In later works by Mrs Hall she did confront members of a higher social class, notably landlords, but at this early stage of her career she did not directly address them. Instead, she told stories that highlighted the difference between good landlords and bad ones. The good ones were those who were resident on their estates and took care of their tenants, the bad ones were absentees and left their affairs in the hands of venal agents. Kate Connor, in the story of that name, was the victim of one such agent, and she journeyed to London to seek justice from her landlord. She found it, because a favourite fictional image of the time was that of the landlord as the power who will save his tenants from the cruelties of the agent. The landlord's chief fault is negligence rather than malice, and the belief among the tenantry (in fiction) is that all will be well if the landlord learns the true state of things and wakes up to his responsibilities. Maria Edgeworth pioneered this image in her novels, *Ennui*, *The Absentee* and *Ormond*, and William Carleton portrayed it in several of his stories. (However, Carleton destroyed it effectively in the moving story of Owen M'Carthy, a man who travelled to Dublin to appeal to his landlord, but who was manhandled and his request ignored.) In Kate Connor's case, the contrite landlord hurried back across the Irish Sea and confronted the villainous agent. Mrs Hall's description of the agent's physical appearance reveals yet again her instinctive snobbery, as well as serving the dramatic purpose of contrasting a wicked, upstart agent with a noble respectful peasant. The agent, Mr O'Brien, had '"his hands in his pockets, his fuzzy red hair sticking out all around his dandy hat, like a burning furze bush and his vulgar ugly face as dirty as if it had not been washed for a month"'.[24] The hard-hearted agent in 'Annie Leslie', Mister Maley, was also described as a low fellow: '"for his mother was a chimney-sweeper that had the luck to marry a decent boy enough, only a little turned three score, and he wore spectacles on his ugly red snub nose"'.[25] In this story too, help comes from above, when the landlord's brother hears of the agent's injustice to the hard-working Leslie family.

The dirt, disorder and discomfort arising from laziness and negligence are so well described by Mrs Hall that a very vivid picture of the indolent Irish peasantry was presented to English readers. The reviewers recognised this straight away. When the sketch 'We'll See About It' was first published in the English annual the *Amulet*, in 1829, a reviewer in the *Atlas* called it 'one of her smart satires upon Irish sloth, negligence and dirt'.[26] In its review of the Second Series of *Sketches of Irish Character*, published in 1831, the *Literary Gazette* remarked on Mrs Hall's concern for the welfare, both moral and physical, of her countrymen, but concentrated on her analysis of their faults:

This is a work which is both delightful and useful; delightful, for the stories are at once interesting and amusing and told with all the life of native vivacity; and useful, for to point out a fault is one step towards correcting it, and many indigenous faults are here truly and kindly touched. That liability to excitement which induces an Irishman to act before he thinks, that procrastinating indolence which insinuates itself under the seductive form of easy good nature, the want of that order which is the great excellence of social life – all these are strikingly painted, and perhaps are the more obvious to every judgment from the simple and natural manner in which they are brought about. The scenes are vivid because they are true.[27]

The *Weekly Dispatch* took much the same view. It praised the series as 'graphic, spirited and original', and went on to say that 'Mrs Hall does ample justice to the warmth of feeling, wit and humour of her countrymen, yet she does not shrink from the task of exposing their faults – the chief of which appears to be a natural indolence, or rather love of procrastination of a most serious nature'.[28]

It is possible, in these early sketches of Mrs Hall's, to detect a half-amused tolerance of native Irish sloth – but her attitude towards the other great flaw seen to be in the national character – intemperance – is ambivalent, in spite of the fact that a connection of hers, the Reverend George Carr, had been one of the founders of a nation-wide campaign against drunkenness some years earlier, anticipating the later total abstinence movement led by Father Theobald Mathew, the Capuchin friar from Cork. Mrs Hall identifies intemperance as 'the national fault' in her story 'Peter the Prophet,' and it is because of over-indulgence in whiskey punch that Corry Howlan, the young hero of that tale, is almost drowned. Whiskey-drinking leads to murder in the story 'Father Mike', when the rivals Phil Waddy and Brian O'Hay, 'partake too freely of the liquid fire'. Yet, in several other sketches whiskey-drinking is mentioned without any authorial condemnation. At Anty McQueen's lavish wedding, described in detail in 'The Bannow Postman', there were glasses, noggins and jugs full of whiskey punch for the guests, and Mr Herriott, that model landlord, provided his tenants with plenty of whiskey on the pattern day described in 'Kelly the Piper'. Poteen-making is the subject of the sketch 'The Fairy of Forth,' and the climax of the story is the confounding, by magical means, of the gauger, or exciseman, who is searching for the illicit still. This is a matter for celebration, not regret, and there is no hint of authorial disapproval anywhere in the sketch. In her later works, however, Mrs Hall's attitude towards intemperance hardened, and she became a crusader against the evils of drink. The lively stories of her early writing career degenerated into tracts, tedious to read, and devoid of any literary merit. Later editions of the *Sketches of Irish Character* often included one of these cautionary tales, 'Good Spirits and Bad', and it makes a melancholy contrast to the sprightliness of the original stories.

Faults, personal or national, are highlighted when the virtues that co-exist with them are noted, and Mrs Hall sought out and presented Irish virtues. The commentators on Irish peasant life had praised the modesty of the women and their virtue and Mrs Hall emphasised this feature in her writings. Her heroines are uniformly chaste and loyal. At times they may be frivolous in speech and capricious in behaviour but they are fundamentally good girls, capable of self-sacrifice. Kate Connor, the eponymous heroine of the sketch, refuses to marry her sweetheart Barney because she did not wish '"to bring poverty to his dwelling"' but continued instead with her plan to seek justice from her landlord in London. Even the outcast informer Black Dennis was finally received back by his wife, whose 'still woman's heart yearned towards its early affection'.[28] Tragedy afflicted many of her heroines. In the sketch 'Captain Andy', Mary Corish rescued her brother James from the barn at Scullabogue but lost her lover, Thomas Garratt, who was on the side of the rebels. Dora Hay's sweetheart, Brian, was killed by her suitor Phil Waddy in the story of 'Father Mike' and Lady Clavis's husband died in a duel in 'The Last of the Line'.

Another type of tragedy afflicted Mary Ryan, who was wrongly suspected of immorality, and in 'Mary Ryan's Daughter', a story that appears in later editions of the *Sketches*, Mrs Hall opens the story with a conversation among the heroine's neighbours that has an authentic ring to it. It is worth quoting in full because it demonstrates the harshness with which moral transgressors were regarded in the peasant society:

"I never saw any beauty in her – that's the truth" – exclaimed one of a group of females, who, lounging around a cottage door, were watching the progress of a young woman toiling slowly up a steep hill, and leading by the hand a very slight child. The cottage was in the valley – and the traveller must have passed the group – for, like the generality of Irish dwellings, it was on the road-side.

"I had the greatest mind in the world to ask her how she had the impudence to wear a bright gold ring on her wedding finger, as if she was an honest woman!" said another.

"And she asking with such mock modesty for a drink of water! I wonder how she relishes water after the fine wines she got used to," suggested a third.

"It was for all the world like a story written in a book," observed the first speaker; "how she left the Uphill farm (as good as seven years, come Easter), and no one ever knew exactly who she left it with – only guessing that it must be one of the sporting squireens, that thronged the country about that time."[29]

Later in the story, Mary stands 'in hopeless anguish uttering no word, shedding no tear – but listening to the vehement abuse poured upon her by her father's wife'. Her step-mother was 'one of a class by no means rare, who imagining that their own virtue is but evinced by condemning, with the utmost violence, every woman who has suffered under the supposition of swerving from the right path'. At the end of the story, Mary is vindicated; she had been legally married and her daughter, Peggy, was acknowledged to be legitimate, a highly important fact in view of the contemporary attitude towards illegitimacy.

Mrs Hall's opinion of Irish peasant womanhood in general may best be summed up in the words of the narrator in one of the early sketches 'Kate Connor', who, speaking of her nurse, the Kate Connor of the title, says 'I bless God that the aristocracy of virtue ... may be found, in all its lustre, in an Irish cabin.'[30] In that story Mrs Hall laid great stress on the bond between employer and servant, and she returned to the theme again and again. Servants might share the Irish vices of sloth, carelessness and intemperance (and in her rôle as a young married woman presiding over her household in London Mrs Hall was well placed to observe these faults) but their fidelity was amazing, and went far beyond the normal bonds of master/servant relationship. In 'Old Frank' in the First Series of *Sketches* Mrs Hall told how the family coachman who had been with them for over forty-two years, saved her grandfather's life during the rebellion of 1798. She recalled that as a child she had

> often heard how he buried the best old wine in the asparagus beds, to save it falling into the hands of the rebels; and how he concealed his favourite horses in the hen and turkey-houses; and how at the risk of his life he carried a forged order to General Roche, who commanded the rebel forces in the town of Wexford; which order purported to come from another rebel chief, and demanded the instant freedom of his master, whose life was thus preserved.[31]

When the time came for the Carr family to leave Bannow for ever Frank would have gone with them to London, abandoning his own family, if it were not for his age. In his own words, he was 'past travelling at eighty-five'.

The fictional servants in the *Sketches* share Old Frank's devotion to his employers; Kathleen Ryley, the peasant girl in 'Mabel O'Neill's Curse', warned her young mistress, Miss Johnson, of a plot to murder the family and burn the house; Martin Finchley saved his master, Father Mike, from the consequences of his folly in treating with the rebels of 1798, and Mary Conway, nurse to Lady Clavis's child, Madelina, revealed to her mistress the details of the scheme the rascally bailiff had devised to ruin the family in the 'Last of the Line'. Most devoted of all was Nurse Keefe, whose former charge, Miss Gertrude Raymond, had been disinherited and whose story was told in the sketch 'Hospitality'. When the one-time nurse heard that her foster-child was in poor circumstances and was working in London as a governess, she gathered all her savings and journeyed there to give the money to Miss Raymond. This journey has a purpose directly opposite to that of Kate Connor, who was seeking help from 'her' family, while Nurse Keefe wished to give it, but each voyage illustrates how strong is the bond between employer and servant, and what perfect trust exists between them.

The Irish were noted abroad for their propensity towards violence and Mrs Hall's stories strengthened this impression. There are very few of them which do not contain some violent incident – murder, treachery, sudden death, abduction, etc. – but in this as in so much else, Mrs Hall's work was to the taste of the times. Readers loved violence and found plenty of it in newspapers and periodicals, not announced in shrieking headlines, but reported in a matter-of-fact way, all gruesome details recorded. Deaths from hydrophobia were described in minute and horrible detail, as were the last agonized moments of victims of domestic accidents such as scalding and burning, while a whole printing industry grew up to satisfy public curiosity about the final hours of executed criminals. Even the most respectable publications catered for this appetite for violence (and for the prurient interest in cases of sexual assault, where the names of victims, even those of young children, were given in full), and reports of violent incidents could be found in unlikely publications. The earliest Dublin-based agricultural magazine, for instance, the *Irish Farmers' Journal*, which was founded in 1812 and lasted until 1827, (and to which Maria Edgeworth was a subscriber) was a specialist publication but its learned articles were interspersed with news items reporting every manner of tragedy throughout Ireland, England, Scotland and Wales.

Reviewers and literary critics seem to have accepted violent action in novels and sketches as normal and unremarkable, so much so that when Mrs Hall's *Sketches of Irish Character* appeared in 1829, the literary comparison that was made was with Miss Mitford, author of *Our Village*, in whose works action is minimal and violence non-existent. Yet one can see why the comparison was made. Mrs Hall's book was not only dedicated to the English author, it was, according to Mrs Hall, 'inspired' by her, and the structure of the works is similar, consisting of a series of sketches in a localised setting. It was a flattering comparison and one which must have gratified Mrs Hall because Miss Mitford was not only a popular author (the first volume of *Our Village* ran into three editions shortly after its publication in 1824) but she had had some critical success. The 'Ettrick Shepherd' (James Hogg) declared his admiration for Miss Mitford in one of the literary discussions in *Blackwood's Magazine*: "I'm just verra fond o' that lassie Mitford. She has an ee like a hawk's that misses nothing, however far off – and yet like a dove's, that sees only what is nearest and dearest, and round about the hame circle and its central nest ... I'm just excessive fond o' Miss Mitford".[32] The critic in *The Times* saw Mrs Hall's work as 'a lively and clever imitation of the ingenious and characteristic volumes called *Village Sketches* by Miss Mitford',[33] but other reviewers spoke of *Sketches of Irish Character* as being on the *model* of Miss Mitford. The *Eclectic Review* called Mrs Hall a 'literary sister' of the English writer, but pointed out that she was 'born on the other side of the Channel, [and drew] from her Irish birth a spirit of romance, an archness of manner, and a tone of pathos, which give to these sketches of an Irish village a character of their own'.[34] The sketches are not an imitation, though 'confessedly suggested' by Miss Mitford's work. The *New Monthly Magazine*, speaking of the Second Series of *Irish Sketches* published in 1831, preferred them to the English ones because the pictures of village life were drawn 'less elaborately and therefore, more naturally, perhaps',[35] than those of Miss Mitford, while the *Literary Gazette* found more of 'interest in narrative'[36] in Mrs Hall's book. The *Dublin Literary Gazette and National Magazine* regretted that Mrs Hall had transferred 'the plan of [Miss Mitford's] *Village Sketches* to Irish scenery and manners. The consequence ... has been that the scope of her book has been too contracted, and instead of being presented with *Sketches of Irish Character* in the extended sense which we might expect, we have only sketches of such character as the peaceful village of Bannow in the County of Wexford and its adjacent neighbourhood afford'.[37] Comparisons of the two works were, in fact, misleading, because although there were some very superficial similarities the books were different in two important respects. Miss Mitford's pleasant, gently humorous sketches had no didactic aim and taught no moral, whereas Mrs Hall was concerned to correct the false views of Ireland that were held in England, and to teach the Irish themselves a better way of behaviour. Most noticeably, the dark seam of violence in Mrs Hall's stories is completely absent from Miss Mitford's sketches of life in 'sunny Berkshire'. The greatest excitement in the *Village* stories is that generated by a cricket match, whereas Mrs Hall's stories deal with some of the wildest aspects of human behaviour. It is true, of course, that many of Mrs Hall's stories were based on her childhood memories of life in County Wexford where the rebellion of 1798 and its horrors were still fresh in people's minds, and that Miss Mitford was living in a quiet and orderly English village where there were no such memories. Yet Miss Mitford, far from living in a rural paradise where everyone carried on the traditions of Old England, was part of a world where everything that had been fixed and settled for centuries was changing under the impact of a new age.

The Industrial Revolution, the Napoleonic Wars and the vast changes in agricultural practice had transformed rural society, and the worker had lost not only his little stretch of common land to the Enclosures, but rights that had been his because of ancient customs. There had been agrarian risings in the eastern counties of England in 1816, in East Anglia in 1822, and there were to be riots all over the east and south of England in 1830, and again in 1834–5. It was desperation that drove farm labourers to burnings, maimings and rioting, the desperation of men who lived close to starvation, often kept alive only by a Poor Law that barely granted them a subsistence level of food and that operated against any hope they might have had of bettering their conditions. In the words of two modern social historians, Hobsbawm and Rudé, 'Pauperism, degradation, desperation and sullen discontent, were ... almost universal'[38] in rural England, yet none of this may be glimpsed in the pages of Miss Mitford. Christoper North in *Blackwood's Magazine* found that her writings were 'pervaded by a genuine rural spirit – the spirit of merry England'[39] and the critic in the *Gentleman's Magazine* enjoyed reading about 'the everyday occurrences invested with true pathos',[40] while the reviewer in the *Athenaeum* admired her works as 'pictures of the manners and feelings of our peasantry'[41] and Allan Cunningham in his 'Biographical and Critical History of the Literature of the Last Fifty Years' in the same periodical approved of her 'hearty sketches of domestic manners, rural pursuits, village pastimes, and her all but living portraits of cottage dames and rustic husbandmen'.[42]

However, a reviewer of Miss Mitford's *Recollections*, writing almost twenty years later in *Blackwood's*, said of *Our Village*, that 'a light something too golden falls uniformly over the picture'[43] and certainly when measured against the reality of English rural life that was true of Miss Mitford's sketches. Contemporary reviewers were aware of the discrepancy between the reality and the fictional world of Miss Mitford but they merely hinted at their doubts, as in the comment of the critic in the *Lady's Magazine* who, reviewing the third volume of *Our Village*, said 'her simplicity may sometimes approximate to the jejeune, and there is a want of deeper shades to create variety and excite emotion'.[44] In a review of two stories by Miss Mitford in the Annual *Friendship's Offering* a writer in the *Athenaeum* pointed out that they were 'of course, clever and agreeable, but written with more than her usual indifference to story and catastrophe'.[45] That could certainly not be said of Mrs Hall, and in one of her short stories set in rural England, 'The Mosspits',[46] her plot hinges on the burning of a rick by a group of farmworkers, evidence that she was aware of the agrarian violence in England.

Politically-motivated violence is an integral part of several of the stories in Mrs Hall's first collection of Irish Sketches and 'Captain Andy' is a tale of the rebellion of 1798 – a true one, according to the author. The story is simple; how a young Protestant girl saved her wounded brother from the barn at Scullabogue by interceding with her Catholic lover, a captain in the rebel militia. When the brother and sister rode away the rebel realized that his sweetheart would never recover from the shock of the Rising, and he set fire to the barn. Later, he fled the country and died in exile, while his sweetheart roamed the country as a madwoman. The story ends with the rebel's body being brought back for secret burial, and the death of the demented girl at his graveside. It is an illustration in fiction of how personal happiness and content are ruined by the violence indulged in by people who think they are fighting in a good cause. When the Protestant boy James Corish joined the Wexford militia, his father sobbed: "It's like spilling one's own blood to fight against one's neighbours", but he blessed his son and advised him: "do your duty, as your father did before you".[47] The possibility that there might be some validity in the rebel cause is not even hinted at in the sketch; the miller who tells his part of the story was a one-time captain with the rebels, and his reason for going to fight was that he "was over-persuaded to join the boys". His wife was against the idea at first, but the thought of being a captain's lady won her over. Mrs Hall's own views are quite clear without being overtly stated; the Protestant boys were upholding the right, in full awareness of what they were doing, while the Catholic rebels blundered into rebellion, without any conviction beyond a desire for excitement, an illustration of the ingrained Irish love of violence for its own sake.

The most violent of all Mrs Hall's early stories is 'Father Mike', the tale of a priest who had had some involvement with the rebel cause. It is not, however, politically-inspired violence that is central to the story, but violence springing from personal passion, lust, and the rivalry between two young men for the hand of a beautiful girl. Murder, an attempted abduction and suicide result, and Mrs Hall's language is suitably extreme. The atmosphere of violence is built up early on in the story – the villain, Phil Waddy, has an eye that is, in unguarded moments, 'fierce and prowling'; his rival, Brian, is 'anxious to engage in a quarrel' and his language is abusive. He could have Phil hanged on the gallows, he says, "and I could make ye a thing that the crow and the raven would turn through the snow from, for sure natur would tell them that even yer corpse was poisoned with the badness o' yer shrivelled heart!".[48] Waddy's eyes 'glared fearfully', and he followed Brian from the public house 'like a bloodhound', laughing 'a low, but fiend-like laugh'. He finally catches up with his victim and bludgeons him with a thick and knotted bough to 'the crackling sound of the crushed skull-bones, and the warm oozing and ourpouring of the red blood, on the fair white robe that covered the earth'.

In 'Black Dennis', it is the after-effects of political violence that are felt, when the wife and child of an informer, along with the man himself, suffer not only cold and hunger, but must live in concealment, isolated from material and spiritual comforts. On his death-bed Black Dennis sends his wife for the priest, and that is how the family's secret is revealed. Good neighbours follow the body to its burial at dead of night, 'far from any other grave', because in spite of their abhorrence of an informer, it would be foolish to "bear spite and hatred to a senseless corpse"'. The author makes a distinction here between the type of informer who reports a crime, and the man who betrays his comrades as did Black Dennis, who

> had been a United Irishman, and one of the most violent order – the projector of more burnings, murders and robberies, than any chief of them all; and when, at last, he found that he could no longer carry on the system of rebellion and plunder, into which he had drawn so many unfortunate victims, he turned king's evidence; many were the men either transported or executed on his statements – all less guilty than himself.[49]

Here again is a portrait of a rebel motivated by nothing more than a love of violence and the hope of personal gain. In 'The Rapparee', in the Second Series of the *Sketches* the central figure is a man who has taken to a life of crime because of a single act of violence. A rich landed proprietor, a Mr Dartforth, who 'had broken his wife's heart by frequent fits of violence', adopted a tenant's son and reared him as his heir, but over-indulged the lad to such an extent that the boy became 'imperious and over-bearing'. In the course of a quarrel between the adoptive father and his son, 'words terminated in blows; Mr Dartforth struck his protegé, and the other, whose tiger-spirit could ill-brook such an insult, hurled his almost-father to the earth'. The daughter of the house separated them and the son expressed his sorrow for what he had done, but Mr Dartforth 'cursed the stripling in the madness of his rage'.[50] The youth left his home and became an outlaw, and the father never recovered from his disappointment.

The consequences of sexual violence are laid before us in 'Mabel O'Neill's Curse', also in the Second Series of the *Sketches*, where a man's seduction and subsequent abandonment of a young girl lead to tragedy many years later, in fulfilment of the curse she lays upon him. 'Mad Mabel' herself, the author of the curse, and instigator of violence, is killed while trying to escape from justice. The seducer's legitimate children die one after another. His natural son is an outlaw, who, unaware of his parentage, tries to burn down his father's house. In 'Jack the Shrimp', the smuggler who struggles with the leader of the excisemen is the father of a young girl who has been seduced and then abandoned by the same man and he is seeking revenge: 'Long and desperate was the struggle – hand to hand, foot to foot – until, as they neared the overhanging edge of the precipitous cliff, the shrimp-gatherer grappled the throat of his adversary; one step more – and both went crashing against the pointed rocks; until the deep heavy splash in the ocean announced that the contest was over'.[51] The seducer was punished but the man who killed him died too, because of his own violent revenge.

These two stories where the plot hinges upon a seduction have a particularly high level of violence, even by Mrs Hall's standards, and one wonders whether it was personal conviction or pandering to popular taste that inspired her to such depths of melodrama. The critics found no fault with her handling of the 'coarser passions' and the reviewer of the Second Series of the *Sketches* writing in the *Athenaeum* went so far as to praise her for 'the good taste, sound sense, and genuine delicate feeling which prompts her to prefer exciting love, and "the milder grief of pity", to stormy ebullitions of power, passion, mysticism and *diablerie* ... .'[52] The lesson of Mrs Hall's stories and sketches, not very subtly taught, is that violence is destructive of all the pleasures of life, and of life itself, and that revenge destroys the avenger as well as his victim. As a fervent Evangelical, Mrs Hall certainly believed this, but she was a shrewd enough writer of entertainments to realise that readers draw vicarious pleasure from violence in fiction while fearing it in themselves and in real life.

Sketches of Irish Life – The Voice of the Colonist

The native Irish attachment to the Roman Catholic religion was widely regarded in England and Scotland as a form of superstition and the priesthood was seen as a sinister force. Priests were suspected of encouraging treason and the behaviour of some individuals during the 1798 rebellion justified these suspicions. The Irish Catholic peasant was seen as an ignorant soul, a ready-made dupe for cunning priests, to whom he avowed absolute obedience and to whom he paid utmost respect. Outsiders believed that a detestation of Protestants united Roman Catholic priests and people, and that relations between the major religious group and the minor one were invariably strained. Mrs Hall attempted to show that not all of these assumptions about religion in Ireland were valid and she emphasised the co-operation that did, in fact, and in some places, and for some time, exist between persons of different religions. The community she described in Bannow was one where religious harmony prevailed. What tensions there were could be traced back to 1798, but most of her characters who had once participated in the uprising, if not ashamed of the part they had played, then at least regretted and were scornful of their youthful folly. The miller in the story 'Captain Andy' did not wish to be reminded of his one-time military rank in the rebel band or of his action in saving his landlord from those who wished to spill Protestant blood. He was, however, still mindful of how the landlord had repaid his debt:

"If it hadn't been for yer good word, thim children o' mine would have had no father. I was ready enough to die for the cause like a man, dacently; but to be hung, jist for nothing, like a dog, was another thing. It'll niver come to that wid me now, God be praised. To be sure, we all have our own notions; but I'll not make or meddle any more, in sich matters, for all the boys wanted to be commanders and gentlemen at once and wouldn't be said or led by their betthers."[53]

Mrs Hall was establishing, for the benefit particularly of her English readers, that her corner of Ireland was calm and contented, and that the passing of time had brought wisdom to one-time Catholic rebels. Father Mike, the priest in the story of the same name, was an exotic and dangerous figure, a Catholic priest who had once been a rebel. Yet, he, too, had learned to regret his past and to repudiate it:

The fact was, [explained the author] that before the Irish reign of terror of 1798, Father Mike, like many of the Romish clergy, had entered into a clandestine correspondence with foreign powers; this had been suspected, and after the rebellion he had been arraigned on the charge of high treason.

Proof, however, was wanting ... [and] Father Mike was acquitted, returned to his parish much wiser than he had left it; and as party feeling died away, the charge was almost forgotten.[54]

Mrs Hall spoke kindly of individual priests; even the once-wayward Father Mike is described as having had a polished education, and is presented as a kindly and tolerant man. His education and priestly training had been acquired on the Continent, and this made him, in Mrs Hall's eyes, superior to the new generation of Catholic priests who were trained in the recently-established college at Maynooth. Her dislike of the Irish-educated clergy is manifest in several stories, and it is very clearly stated in the later guide to Ireland, written in conjunction with Mr Hall. A young priest is a major character in the sketch 'Lilly O'Brien' and the author presents him in a respectful light, observing that he was 'a most extraordinary Irishman; cautious and prudent, even when a youth, and gentle and constant'. He was 'useful in his ministry, for he had kindly feelings towards all his fellow creatures'.[55] The priest who tends the dying woman in the story of Luke O'Brien is a 'kind and benevolent man' who 'hastened to his duty' of administering the last rites, and Father Connor in the sketch 'Black Dennis' is even more kind and charitable for he takes an orphan boy into his home. The warm relations between pastors of different faiths that Mrs Hall had known in her childhood was stressed in the sketch 'The Bannow Postman', when the parish priest and the minister had a collection in their respective churches to aid a woman in poor circumstances. They could also be seen 'conversing at the door [of her cabin] as to the best means of procuring the industrious woman continuing employment'[56], and at Anty McQueen's wedding parish priest and minister sat side by side. However, a clash between priest and parson is described in the story 'Lilly O'Brien'. A young wife has died in childbirth and there is 'a long and loud debate between the Protestant and Catholic priests as to who was to perform the last rites' because although the young woman was a Protestant she had been married to a Catholic. The widower awoke from his grief and said:

"Plase your reverences, I'm a Catholic, and ever was and will be; but she that's gone from me was born a Protestant – married a Protestant – and as she died one, so shall she be buried, and that's enough; and what's more, I promised her, when I didn't think that death and desolation would come at this time, that if the child was a girl it should go wid' her, if a boy, wid' me."[57]

The young man is referring to the accommodation for 'mixed'

marriages made by the Irish churches, both Roman Catholic and Church of Ireland. The agreement was as he described it, that male children of the union should follow the religion of their father, and females that of their mother. This benign arrangement continued up to the early 1820s when the Evangelical spirit sweeping through the Irish Protestant church changed everything. Mrs Hall was remembering an era of religious toleration that had already passed away, and in her later works she was forced to take account of changed circumstances.

The Irish peasant was commonly held to believe in fairies – a superstition to be sure, but an attractive one, and the begetter of many good stories. Tales of the supernatural, the 'Gothic' novels of the 1790s, were still popular in the early years of the nineteenth century so there was a willingness on the part of the reading public to accept stories about spirits. Allied to this taste was the interest in primitivism and the early manifestations of peasant cultures. Traditions, beliefs and legends, all that makes up folk knowledge, or folklore (a term not in use until 1846) were popular among the educated classes, and Irish folklore was a particularly rich field. Thomas Crofton Croker's three-volume collection of *Fairy Tales and Traditions of the South of Ireland*, published between 1825 and 1828, was followed by Samuel Lover's *Legends and Stories of Ireland* in 1831 and his *Popular Tales and Legends of the Irish Peasantry* in 1834. Irish legends also appeared in accounts of tours, such as those of Cesar Otway, the editor of the Irish *Christian Examiner*, and were sprinkled throughout Irish novels, notably those of Lady Morgan. It was natural, then, for Mrs Hall to include local stories within her sketches, as in 'Old Frank', where the coachman's story of fairy trickery is remembered by the author and re-told for her readers. She describes the old coachman as having had 'a most confirmed belief in banshees, cluricawns, fairies and mermaids' but adds that the old man's daughter was sceptical. Another folk tale in its entirety is in the sketch 'The Fairy of Forth', which is also narrated by an acquaintance, an old man living on the mountainside of Forth. Both of these stories have the ring of authentic folklore, but two others are of dubious provenance. 'Take it Easy', which is an older woman's advice to a young bride, is the telling of a fairy story to point a moral but it suffers from a sentimentality and an excess of whimsy which is not found in a true folk tale. The same criticism applies to the story within a story found in 'The Bannow Postman'. Grey Lambert, the old eccentric, tells Anty McQueen, the girl who is about to be married, a little tale about fairies and flowers which is, in reality, a piece of advice about love and fidelity. The language of both these tales is demotic, but they do not ring true, and it is Mrs Hall's sentiments we hear, not the expressions of ancient beliefs.

In common with other Irish writers of the period (notably William Carleton) Mrs Hall attempted to convey the flavour of Irish peasant speech by the use of dialect. Her native characters speak an English that is distorted; 'beyond' becomes 'beyont', 'moonbeams' became 'moombames', 'please' became 'plase', 'neat' is 'nate', 'rent' is 'rint', 'plenty' is 'plinty', and so forth. 'Ye' is used for 'you', 'yer' for 'your', 'wud' for 'would', 'bud' for 'but', and 't' for 'th', as in 'tunder' and 'wit' and vice versa when 'th' is used for 't' in 'throuble'. The speech is not entirely consistent and varies from character to character without, however, differentiating their personalities. In this Mrs Hall is unlike William Carleton who uses dialect more subtly and with greater understanding of the people whom he is portraying. This, almost more than any other feature of their writing, illustrates the gap between the benevolent outsider looking in, as Mrs Hall does, and Carleton, the insider, speaking from a peasant heart.

Mrs Hall, nevertheless, did have some success in catching the rhythms and the flavour of Irish native speech. Her beggars, servants, labourers, and pedlars use language rich in simile and metaphor and full of vivid imagery. Peggy the Fisher, the travelling saleswoman in the sketch 'Lilly O'Brien' describes the heroine as 'withering with sorrow' and fading away 'like the mist up the mountain', and more comically, when Lilly falls through the thatch of an illicit distillery as 'tumbling ... through the black roof like a snowball'. Lame Larry, a mendicant visitor to the inn in 'The Rapparee' taunts a companion who has been boasting of his wisdom with the words '"What a wonder that is, to be sure ... as if you were one who could shoe the goslins, catch a weasel asleep, or spit a sunbame"'.[58] Alick the Traveller who played an important part in the love story of Annie Leslie advised the girl's unlucky suitor that she was in love with another, saying '"Yarra a much sense ye have, or ye wouldn't be looking after empty nuts – what the divil would be the good o' the hand o' that cratur widout her heart?"'[59] Kate Connor, who had been evicted by the landlord's agent, cursed him, saying '"sure as there's light in heaven, in his garden the nettle and the hemlock will soon grow, in place of the sweet roses"'.[60] and Mary Clavery (of 'The Bannow Postman') in similar case, remembers with pride that, in spite of her family's troubles, '"no blight was on our name, nor isn't to this day – thank God for it! – for nobody breathing can say, 'Thomas, or Mary Clavery, ye owe me the value of a thraneen"'.[61] The mad-woman, she was not admitted immediately to a farmer's house poured forth invective:

"And that's yer fine breeding, is it, Katy Ryley? – to stand staring at an aged woman *outside* the door-cheek! – at one whose head is grey – whose feet are sore – whose lips are dry – whose bag is empty – who has neither kin nor friend near, to say 'God save ye' – nor a stick or a stone to set her mark upon – where she may lay down her bones and die?"[62]

This vigorous language is in sharp contrast to the stilted phrases that are used by the more refined characters in the sketches and by the author/narrator, whose occasional comments almost always strike a jarring note, even when she is not overtly moralising. Beggars, wandering men, or 'gaberlunzies', as she sometimes calls them, are among her favourite characters, and one in particular, Denny Dacey, who plays a key role in solving the mystery of Mary Ryan's daughter, describes with humour, pathos, and dignity the difficulties of his life. Ladies and gentlemen, he advises, should be 'handled like a nest of young thrushes' if they are to part with money. Flattery is essential, but it must be wisely chosen. A blessing on the beauty of women will smooth their frowns away but he recalls the mistake a young woman made when she held up her baby to a frosty-faced old lady and bade her remember her own family. That 'turned her to hard vinegar' for she had none. Ladies accompanied by their children were easy to flatter. All one had to do was to 'praise the children; and if they're as ugly as frogs, lay on them all kinds of angels; and if they are roaring wicked with ill-temper, call them little lambs'. Beggary, he adds, is not a matter of choice, but of necessity, in order to avoid starvation. 'I was so thin', he recalls, 'one or two of the hard summers, that if it wasn't that I had the wit to put stones in my wallet, I'd have been blown away'.[63] The author, in a typically stilted paragraph, comments:

I wish I had space to recount all Daddy Denny's stories. Some of them could not fail to make you weep; and his transitions from humour to pathos were truly characteristic of his calling. There are many who cannot fail to remember this energetic yet lazy personage, who latterly begged from habit; and who was at all times trusty, and trusted by many of his superiors ...[64]

There is no condescension in Mrs Hall's reporting of peasant speech and it reads as an honest attempt to convey to others the pleasure she herself took in it, but the same cannot be said of her rendering of peasant writing. In the two series of *Sketches* there are three letters purportedly written by peasant characters and nothing about them seems authentic. They are not semi-literate, as one might reasonably expect from uneducated young people, or awkwardly phrased as would be the case when the correspondent is unused to writing, but are cleverly composed to illustrate how Irish peasants mangle the language. There are no errors of spelling, but the sounds of English as spoken by Irish peasants are used: 'bateing' for 'beating', 'taze' for 'tease', 'sowl' for 'soul', 'minit', 'minute', and so forth. These letters are obviously meant to be laughed at by an educated English reading public.

The assumption in both series of *Sketches* is that the Irish peasant is inferior to an Englishman – any Englishman – and those stories that highlight his and her faults and idiosyncracies are illustrations of this belief. It is not very subtly done, and in several of the sketches the belief in the essential superiority of everything English is openly stated. Annie Leslie, in the sketch of that name, had a 'fair skin', and 'small delicate mouth [that] told of English descent'. Her father was indeed an Englishman, living in Ireland, who showed his landlord the improvements that he 'as an English farmer, thought might be profitable to the estate', and who possessed the 'natural independence' of an English yeoman'.[65] Mrs Cassidy, the well-to-do widow in 'Lilly O'Brien', who kept a clean, well-furnished house, 'might, from her habits, have passed for an English woman'.[66] Mark Connor, the farmer and pig-dealer, hero of 'The Wooing and the Wedding' was proud to announce that he had an English wife and aimed to make his home the 'very moral of an English cottage'. His wife Helen, caused consternation in the neighbourhood by her progressive English ways, but her young brother-in-law, Matty, was quick to appreciate her superiority and 'improved much by the wise precepts and uniformly good advice of his new sister'.[67] He accepted that the English people were 'finer-like' than the Irish, and Mark himself wished that Helen's sister would come over and teach them how to fatten pigs her way. This sketch concludes with an overt statement of what must be accepted as Mrs Hall's views of those she calls, from time to time, her countrymen and women. The dialogue between the English couple, Helen and her father concerns the character of the Irish. The father comments

"Really, the people are so careless"

"Yet good-natured," said Helen, smiling

"So insincere."

"Not so, father, they always *mean* to perform what they promise; but they are, I confess, too apt to promise beyond their *means*."

"So passionate."

"But so forgiving."

"So extravagant."
"So very hospitable"
"So averse to English settlers"
"About as much as we are to Irish ones"
"Averse to improvement, then."
"Not when convinced in what improvement consists."

"Helen, do you know it is very hard to convince an Irishman; he has so many quips and cranks, and puzzling sayings, and would prefer being reduced to expedient, to attaining anything by straightforward means – provided it was not too troublesome."[68]

These views of Mrs Hall's never really changed, and they infuse all her Irish stories and sketches. In her later work, especially when she had gone beyond mere reminiscence, she broadened her view of Irish problems, social and political, and went so far as to put forward suggestions as to how they might be solved, but she never saw the Irish people, particularly the Irish peasants, from any viewpoint but that of the civilizing colonizer among primitive yet lovable natives.

FIVE

Lights and Shadows – a melancholy book

The failure of Westley and Davis, the publishers, in 1837 which caused financial disaster for the Halls also meant that Mrs Hall needed a new publisher for her Irish stories. Her work was in demand in many of the periodicals, and the obvious progression was, as before, a collection of her stories and sketches. In fact, there was no problem, because the owner of the *New Monthly Magazine* in which so many of her sketches had appeared between 1834 and 1835 was Henry Colburn, probably the most successful, and certainly the most energetic publisher in England or Scotland. Colburn's career is of interest; he originally published Gothic romances and French novels, and was the publisher of Lady Morgan's *O'Donnell* in 1814. His most lucrative and successful publications, however, were what were derisively known as 'the silver fork' novels. These were novels of fashionable life, and the term 'silver fork' derives from an essay of William Hazlitt's in the *Examiner* in 1827. In the essay he mocked the novelist Theodore Hook for what Hazlitt called his pre-occupation with the silver forks used by higher-class persons when eating their fish. Hook was one of many novelists who described in detail the lives of well-born and wealthy characters, supposedly fictional, but in many cases recognisably based on prominent living personages in high society. Politicians, hostesses, dandies, dangerous beauties, all appear in novels by writers who included Marguerite, Countess of Blessington, Edward Bulwer, Lady Charlotte Bury, Benjamin Disraeli, Mrs Gore, Lady Caroline Lamb, Letitia Landon, T.H. Lister, William Maginn, Lady Morgan, Caroline Norton and Robert Plumer Ward. These novels were all wildly popular and sold well in spite of their expense – one and a half guineas being the usual price for a three-volume set. Henry Colburn was the publisher of the majority of the novels and they were well produced, with good bindings, clearly spaced print lines and wide margins. Reprints were started in 1831 in 'Bentley's Standard Novels and Romances' series (Colburn having joined with Richard Bentley in 1830 to form the firm of Colburn and Bentley), and in

'Colburn's Modern Standard Novels' in 1835. These reprints were in one volume, with closer, smaller print and were significantly cheaper.

Colburn was a pioneer of the public-relations industry, and his authors never lacked favourable publicity. He had three magazines under his influence – the *Literary Gazette*, which he had founded in 1817, the *Athenaeum* which had been founded in 1818, and in which he had a half share, and his original periodical, the *New Monthly Magazine* which he had started with Frederick Schoberl in 1814. Through the columns of these magazines he promoted his authors' interests, not, as other, more gentlemanly publishers did, with simple announcements of forthcoming works, but by more dubious methods. Alison Adburgham, author of a study of the fashionable novel, *Silver Fork Society*, has described his activities:

> He commissioned lengthy reviews by writers upon whom he could rely to review the books favourably. Sometimes the author himself wrote the review. And he tickled the readers' curiosity by editorial paragraphs about forthcoming novels 'of outstanding interest to high society', hinting at the aristocratic position of their anonymous authors, and surmising that the characters in the books were based upon well known persons. Subtly he ensured that a forthcoming novel would be a subject of conversation at dinner parties, on the clubs and in the drawing rooms, before it ever appeared in the bookshops.[1]

Mrs Hall's sketches of Irish life did not, of course, fall into the category of 'fashionable' novels, but the first and second series of these sketches had sold well in 1829 and in 1831, and the individual articles in the *New Monthly Magazine* had proved popular, so Colburn was taking no risk. Henry Colburn, in fact, in spite of his taste for frivolous literature and his dubious methods of advertising, was a man with an eye for quality. He was the first publisher of Pepys's diary in 1817, the first publisher of John Evelyn's diaries in 1818, and in 1825 he became the first publisher of *Burke's Peerage*. Mrs Hall's sketches were hardly likely to rank with those publications, but they were a worthy contrast to the fashionable novels, and in any event Colburn had already published Mrs Hall's 'English' novels, *The Buccaneer* in 1832, *The Outlaw* in 1835, and *Uncle Horace* in 1837. They were not in the 'fashionable novel' mode; *The Buccaneer* and *The Outlaw* were historical romances, and *Uncle Horace*, although set in contemporary England was a story of middle-class life, with a high moral tone.

Lights and Shadows of Irish Life,[2] the title given to the new collection of Mrs Hall's work which appeared in 1838, was not of her own choosing, as she explained in the Introduction. It had been suggested by her publisher and she had adopted it 'with reluctance, as too nearly resembling that which belongs to one of the most exquisite volumes of modern times'. This was *Lights and Shadows of Scottish Life* by John Wilson, the Professor of Moral Philosophy in Edinburgh University, famous as 'Christopher North', contributor to the dialogues *Noctes Ambrosianae* in *Blackwood's Magazine*, and a widely-respected author. It was a collection of tales about simple folk in Scotland, with a strong moral message, and the stories had originally appeared in *Blackwood's Magazine* before their collective publication in 1822. Colburn saw the resemblance and decided to capitalise on it by the use of a similar title for Mrs Hall's book. She described herself as

following, though at a respectful distance, the example of the eloquent and distinguished Scotchman, and endeavouring to do for my country what he has done for his – to make the character of the Irish more extensively known and better understood, to excite a generous sympathy for their sufferings, a kind indulgence towards their faults, and a just appreciation of their virtues.[3]

Lights and Shadows is in three volumes, and Mrs Hall, in the Introduction to the work, adverts to the fact that several of the stories in the second and third volumes had already been published. She says 'they have already courted favour in Periodical Works conducted by my husband', (the *Amulet, Spirit and Manners of the Age* and the *New Monthly Magazine*), thereby giving Samuel Carter Hall some credit and exhibiting a wifely demureness, and goes on 'I hope I may consider myself justified in collecting them; and that, in their present form, they will not be deemed an unworthy contribution to a class of literature which is designed to carry information while affording amusement'.[4]

The first volume of *Lights and Shadows* consists in its entirety of a long story entitled 'The Groves of Blarney', which is also the title of an Irish song extremely popular at the time. Several of the characters, notably a half-witted girl called Aileen, sing in the course of the story, and the songs are described as old Irish airs or 'Carolan's sweet melodies'. The author notes in a foreword that 'the songs have been set to music by a Mr Alexander D. Roche, and have been published by a London company, Messrs Duff and Company, of Oxford Street'. The presence of so many songs helped Mrs Hall later to transform the story 'The Groves of Blarney' into a play with the same title, and this was possibly her intention from the start.

According to Mrs Hall, the story was based on an incident that occurred in Blarney, Co. Cork, in 1812 and she pays tribute to a fellow-author, who shared her concern that Ireland should be better-known abroad, saying

> For the history and character of the place [Blarney], I refer them to the details of my friend Mr Crofton Croker – to whom Ireland is so largely indebted. He has employed rare talents and industry in her cause; and was among the first to direct the attention of England to the vast stores which she possesses – stories from which profit, information and amusement may be largely drawn.[5]

(There was a close friendship between the Halls and Crofton Croker, and it was Croker who proposed Hall for membership of the Society of Antiquaries in 1842.) The tale is a highly melodramatic one of kidnapping and revenge, and the role of villain is played by an improbable character called the Griffin, who is a species of land-based female pirate. She is insulted by the hero, and gets her revenge on him by tricking him into breaking the vow of temperance he has made to his fiançée, a young widow. The Griffin and her wicked associates then kidnap the widow's little son, but the hero rescues the boy, the widow forgives him for breaking his vow, the Griffin is killed and all ends happily. Virtue of a special kind is rewarded in this story because the hero, Connor Gorman, though legitimately angry with the woman who has tricked him and lost him his fiançée's respect, shows her no violence, and it is her own wickedness that leads to her death in a fall from a cliff top. Connor, however, exemplifies some of those faults to which the Irish people were seen to be prone. It is because he drinks and becomes quarrelsome that his fiançée, Margaret, exacts from him the vow of temperance. Mrs Hall comments several times on the damage done to Irish men and women by their love of drink, 'the darkest and deepest curse under which Ireland labours – the enemy they put into their mouths to take away their senses'.[6] Whiskey is especially dangerous, and 'more than half the murders in Ireland are perpetrated under the excitement produced by ... that accursed beverage'. Mrs Hall suggests an alternative drink for the peasantry, hoping for the

> introduction of a quieting and not a maddening draught [which] would go a great way in subduing the riots and dissensions which disgrace the annals of my country. I would recommend one – Guinness's porter – in the words in which it was recommended to me by a genuine Irish peasant – "That's the stuff that makes the arm strong and the heart stout, without being savage".[7]

No question here of total abstinence, such as was later preached by

Father Mathew, for whom both Mr and Mrs Hall had a great admiration.

Connor's sister, Alice, also displayed some of the faults Mrs Hall considered to be peculiarly Irish. She was slovenly in her ways – the hens were not fed from a trough, but ate from a saucepan, and the pigs were not properly fastened into their sty but were allowed to roam freely, and she was inclined to put off until tomorrow what should be done today. She could not, for example, understand why the haymaking had to be done at once to take advantage of fine weather and she was scornful of Margaret Lee and her proper 'English ways'. Margaret, significantly, was the daughter of 'an intelligent and industrious Englishman' who had been brought over by a local landowner to 'superintend his farms, cultivate his lands, and watch over his hot- and greenhouses'. Naturally, she had grown up with an outlook that was more English than Irish, and she disapproved not only of Irish drunkenness and belligerence, but impetuosity, carelessness and improvidence. A servant girl's request for a few shillings to enable her to get married draws from Margaret the exclamation 'A wedding, and nothing to begin with; children and nothing to give them; premature old age; a broken heart, and a narrow coffin'. This is Mrs Hall speaking through Margaret, and her voice is heard again in the reply made by Margaret's sister, Flora, though it lacks the music of the first outburst:

"And yet", said Flora, "how warm, how true, how affectionate they are to each other! I have seen grown men and women starving, absolutely starving, carry their decrepid [sic] parents on their shoulders, and share with them the scanty morsels which their poverty wrested from the poverty of others. Peasants' wives, however beautiful, and in early youth how beautiful they are! are never known to break the marriage vow, which ties them to poverty, and often to harsh, unfeeling husbands, for that horrid whiskey at times renders them more than half-mad."[8]

The author makes it quite clear that English ways are best: Connor's parlour, though 'excellent in the house of an Irish bachelor farmer, would have been little thought of in an English cottage', and the addition of a back door to his house was an English idea (and it never appealed to his slovenly sister). Margaret's parlour was what one would expect to find in England, and she had fashioned a bower in her garden, which she had 'trained with English skill'.

Colonial attitudes predominate in this melodrama and yet it contains a passionate plea for English understanding of the Irish character which is more focused, more articulate and more thoughtful than heretofore. Mrs Hall calls on the English reader to accept what she has to say as a truthful observer:

Ah, trust one, who though she loves the land has never yet written of it a *line of false praise*, but has freely and honestly censured what she hoped to amend – believe her when she tells you not to credit all the evil that is reported of a country which during a long lapse of years has sent forth so many of the bold, the brave, the brilliant, the beautiful, the glorious to rank foremost among the admired and celebrated of the world!⁹

In her own evaluation the writer is to be trusted because she knows Ireland so well that she has no prejudices; she can see her countrymen and women clearly, and can assess their faults as well as their virtues. She is by now very publicly pressing the case for sympathetic understanding of her country, and the unattractive Irishmen and women abroad are forgotten, and attention is focused on those who were a credit to their country when they emigrated – 'the brave, the brilliant, the beautiful, the glorious'. This passage, balancing the earlier criticism of Irish faults, with its appreciation of Irish virtues is typical of Mrs Hall's efforts to be fair-minded about the Irish character. It is not enough to illustrate generosity, kindness, fidelity and purity by means of an anecdote, and allow the facts to speak for themselves, Mrs Hall must intrude with an authorial comment (even if, as here, it is put in the mouth of a fictional character) to underline what has been shown. By this means she establishes her position as impartial observer, yet true friend. This appeal to readers was not repeated in the stage version of 'The Groves of Blarney', which starred the popular Irish actor Tyrone Power as Connor Gorman, and which ran for a season at the Adelphi Theatre in London in 1838. The play is very different from the book – the emphasis being on its melodramatic and farcical elements, and there are no appeals to anything other than mere sensationalism. It was not received very kindly by the theatre critics, but there was general praise for the 'Groves of Blarney' in its written version. The critic in the *Spectator* considered it to be the only story in the three-volume *Lights and Shadows* 'of sufficient length, purpose, or novelty to require more criticism than is implied in a general recommendation'.¹⁰ He did, however, comment that there was 'an occasional disposition to thrust the opinions of the writer unnecessarily upon the reader'. This rare criticism of Mrs Hall's didactic methods may have referred to what he called 'the drift of the story', which was to point out various faults of Irish character, including a love of alcohol and a general recklessness, or it may have been inspired by the attacks on English attitudes towards Ireland which are scattered throughout the story.

These take the form of speeches addressed to an audience and sit uneasily into the narrative. For instance, the heroine, the young widow, Margaret Lee, when discussing with her younger sister, Flora, the forthcoming visit of their English relative, Peter Swan, who is going to write a book about Ireland, delivers a monologue about insensitive English travellers: '"Persons some of them almost as self-conceited and as ignorant as our Cockney cousin come over here, and go back exclaiming against the wretchedness, misery, starvation and madness of a people whose habits they cannot understand; and yet have neither intellect nor feeling to devise means for improvement of a glorious and suffering nation"'.[11] The character of Peter Swan, the tourist and would-be writer, was thought by some reviewers to be exaggerated, 'a gross caricature' according to the *Spectator*, and in the view of the *Athenaeum* 'an excrescence upon a natural story, so utterly out of nature ... no business in a picture professing to offer a faithful portraiture of modern life and manners'.[12] He is indeed overdrawn, designed to provide comic relief, and it is easy to see how the character could successfully be transferred to the stage and gather many laughs. He is, in fact the 'stage-Englishman', a neat reversal of the traditional Irish/English rôles.

When first seen, Peter Swan is wearing a shawl knotted around his shoulders (an Irish June is like an English April, according to Mrs Hall), and he is 'so meek, so pale, so overloaded with this world's superfluities, and withal so lachrymose' that he is in direct contrast to the 'merry ragged peasants' he meets in Ireland. His adventures are predictable; he says the wrong things in a peasant gathering, is ducked in a bog-hole, is caught by outlaws, has his head shaved and his face blackened, and finally goes home believing that he has been bound by oath to be a United Irishman for the rest of his life. He speaks a strange form of English, presumably Mrs Hall's idea of Cockney dialect, with 'h' interposed in front of vowels, as in '*h*uncivilised, *h*unnatural, wild *H*irish monster', yet he does not drop the aspirate but speaks of 'home', 'heaven', 'husband', 'hat', and so forth. His ignorance of Irish matters is total, exemplified in his very un-Irish action of calling in the police when Margaret's little son is kidnapped. He 'talked of Bowstreet officers, and city police, and "the proper authorities"', and demonstrated 'that sort of virtuous indignation which the English always feel at a maladministration of the Irish laws'.[13]

Apart from his lending broad humour to a highly melodramatic tale Peter fills the useful function of providing other characters with an opportunity to speak their mind about the failings of English tourists, and the inaccuracy of some guide books to Ireland, and finally, the injustices of English rule in England. In one of the few natural-seeming dialogues, when the absent Connor Gorman is mentioned, Peter describes him as a '"genuine specimen of the race ... one degree and a quarter removed from positive cannibalism"', drawing from Margaret the exclamation '"How dare you put such stuff into your foolish book?"'. She goes on: '"After this fashion is the country injured and insulted. Petty scribblers make a month's visit, scrawl, and then print their nonsense"'.[14] Another type of tourist would be welcome; a visitor who came without preconceived ideas, with sympathy and with a willingness to learn. '"I should like the English to come and see us"', says Connor, so that they can '"judge for themselves if we are the despicable set of savages we are represented to be; our greatest crime is *our poverty* – but, to be sure, *that* is a crime all over the world"'.[15] The message here is that English writers who produce books about Ireland cannot be trusted and as it is not possible that English people *en masse* will come to 'Ireland to see matters for themselves', the only way of learning the truth about Ireland is to listen to an Irishwoman who has already proved in her writings that she can be trusted not to show prejudice.

That is Mrs Hall's estimate of herself, unaware that she was speaking as a writer whose English values have already prejudiced her against those very people she thinks she is defending. Her colonial attitudes and class prejudices had already been revealed in small but significant ways in the *Sketches* and her latest work was no different. The reviewer in the *Dublin University Magazine* commented on this aspect of Mrs Hall's writing when he noted that 'from her residence in England, Mrs Hall has acquired a habit of contrasting the conduct and opinions of the English and Irish. The result of the contrast is sometimes not very favourable to her own country; but the reproof is conveyed in such a spirit of kindness, that it is impossible to regard it as otherwise than as the advice of a friend'.[16] There is a basic uncertainty in Mrs Hall's vision of herself, and it is reflected in the character of Margaret, the heroine of 'The Groves of Blarney', who although Irish-born, is of English parentage, and follows English ways. Mrs Hall at all times proclaimed herself Irish and frequently referred to the Irish people as 'her' countrymen and women. She was the interpreter of their ways and customs, and *Lights and Shadows* is, more so than the *Sketches*, an explanatory book, addressed to 'my English readers', who may be surprised to learn that there are well-kept cottage gardens in Ireland, that every Irish hunt has an 'attendant fool', that

Irish parlours can be neat and well-kept, and that 'the north of Ireland is a trading and consequently a prosperous part of the country'.[17] In that rôle of a teacher she was, of course, Irish, but now and then she changed her stance, even while using her authorial voice, and spoke as an Englishwoman; for example, when describing Irish pride of family, she declared: 'How ridiculous it would appear to us in England, to hear a tradesman expatiating on his connection with the aristocracy'.[18] Brooding on the wretched conditions in which the peasants lived, she noted that they had 'worse food than we in rich and happy England bestow upon our dogs',[19] and, in contemplating the effect of alcohol on the Irish constitution, she stated: 'when Irish spirits are raised, we at this sober side of the herring pond have little idea of the height they mount to',[20] indications that she thought of herself, in respect of behaviour, as English and so could not avoid the note of condescension in her voice. This attitude changes, however, when she wishes to plead Ireland's case against English maladministration and then she is fully Irish, as when she speaks through a scholarly Irish character, Marcus Roche, in 'The Groves of Blarney'. He brings the story to an end with an emotional speech that would have formed a fine climax to the dramatised version when he asks

"Why have we been suffered to drag on our existence like a poor relation – instead of a cherished sister? It is in vain that England boasts her sympathy with our distress, while our peasants continue to starve; it is absurd to talk to a man of peace and contentment, while his children are dying of hunger, in nakedness, by the wayside. High-minded English individuals have ever been liberal to us; but we would prefer English justice to English charity. There is abundant scope for speculation in the island, where English capital might be employed to the advantage of both countries."[21]

The second volume of *Lights and Shadows* contains a selection of stories and sketches most of which had already, as Mrs Hall had pointed out in her Introduction, appeared in print. They followed the pattern seen in the *Sketches* – some vivid passages of peasant speech, followed by more restrained authorial comment, easy and skilful narrative, simply told, and anecdotes illustrating various facets of the Irish character. Both faults and virtues are highlighted – drunkenness, impulsiveness, belligerence, carelessness, superstition and laziness, all these are counterbalanced by instances of Irish peasant fidelity, shrewdness, genuine religious feeling, courage, kindness, female purity, hospitality and filial devotion. Irish pride, the pride based on false values, is a major Irish fault, and it is found in all classes of society. In the *Sketches* Mrs Hall had been scornful of the 'half-gentleman', a type she called 'a noxious species, almost peculiar to Ireland'. The young men were idle, full of pride of family, and 'draw corks, tell lies, smuggle occasionally, thrash bailiffs, seduce innocent girls and end their lives generally ... either in New South Wales, or in a jail'.[22] She expanded this theme in the long short story 'Harry O'Reardon: or Illustrations of Irish Pride', published in Philadelphia in 1836, and later forming part of the third volume of *Lights and Shadows*. It is an examination of the consequences of foolish pride. The central character is a young man, penniless, but too proud of his family ancestry to take up menial employment. His adventures take him to Dublin, to Liverpool and finally to London. At every step of the way he is hindered by his false pride and is unable to profit by the help given to him by benefactors, and at last he drifts into crime and dies in prison. Mrs Hall prefaces this tale with the comment: 'The pride of ancestry may deserve to be considered a noble pride when it stimulates to exertion, and animates to virtue. But, unhappily, in Ireland it rises trumpet-tongued against every species of employment derogatory to the memories of the O'Connors, the O'Reardons, MacMurroghs, MacCarthys, O'Briens, or O'Tooles'.[23] She adds 'A bushel of Irish pride is not worth a grain of English independence'. True to her usual practice, however, Mrs Hall highlighted an Irish virtue in this story. This was the fidelity shown by Harry's foster-sister, Moyna Roden who although 'below him in birth' remained devoted to him, to death and beyond.

Lights and Shadows so, is in many ways, a continuation of the *Sketches of Irish Character*, but in one of the stories 'The Dispensation' Mrs Hall shows a change in her attitude towards Irish Roman Catholic priests. In the earlier work she had been respectful towards, and understanding of the relationship between priests and people. As a child in Wexford she had many contacts with priests and friars while she lived in her step-grandfather's house – in the Introduction to the Fifth Edition of the *Sketches* in 1854 she describes the visits made by Roman Catholic clergy to the house in Bannow and the cordial relations between them and the family – and according to a modern writer, Father Butler, O.S.A., her grandmother was a Roman Catholic.[24] Unfortunately he gives no grounds for this assertion, and it is unlikely to be true, as Mr Hall in his memoir states unequivocally that the grandmother was of Huguenot descent, and Mr Hall had a keen eye for the particularities of organised religion. Nevertheless, there obviously was a family tradition of tolerance, and this was reflected in Mrs

Hall's earlier work. The priests described there were courtly men, educated on the Continent, as were the majority of those who visited the Carr household in Bannow. In 'The Dispensation', however, Mrs Hall recalls that there was one priestly visitor for whom she had had no respect – one of the new men, educated at home in the new College at Maynooth, thrusting and uncouth. On her memories she based the character of Father Neddy Cormack, the villain of the story.

Mrs Hall's language in this sketch of a priest is unexpectedly intemperate and the effect she achieves is one of caricature. Father Neddy wishes his nephew to marry a well-to-do girl of the parish, and to this end refuses the girl a dispensation to marry her first cousin with whom she is in love. The priest's appearance indicates his faults of character: 'His eyes told you that if the creature possessed power in proportion to cunning, it would indeed be fierce and dangerous. The thing would have made an admirable attorney but a bad counsellor, and certainly was a very unfit director of the spiritual or temporal affairs of the parish, which he endeavoured to *rule* – not *guide*'.[25] The use of the words 'creature' and 'thing' reduces the man to animal level, a tendency of Mrs Hall's when speaking of Irish peasants, and it is a measure of her disdain for the new native priests. Father Neddy's speech is not that of an educated man; no scraps of French or shreds of Latin, but coarse and sprinkled with unlikely oaths: '"I'll excommunicate them all ... by Saint Peter and Saint Ambrose and Saint Obadiah"'. He could advise his nephew on how to make love to a young woman: '"D'ye think I've been hearing confessions from all manner of faymales for the last forty years, without knowing how to manage 'em?"'.[26] This was exactly what the majority of Protestants wanted to hear, confirmation from a priest's own lips, as it were, of what they had always suspected, that priests had a sexual rôle in the lives of their female parishioners. It was still difficult to define that rôle exactly, but the image of a woman alone in the dark with a man, whispering intimate details of her life into his eager ears was a very potent one. It may even have accounted for the success of the story in popular and critical terms. ('Mrs Hall's best story to date',[27] was the view of the critic in *John Bull.*) Another reason was the pandering to the public perception of the priest as money-grabber, a man whose faith shrivelled at the sight of gold. Catholicism was reduced from a belief to a commodity, a sure sign that it was not a true faith at all and that the Reformation had been justified. Rather cleverly, and with more subtlety than she had hitherto used in the tale, Mrs Hall put this view in the mouth of the young Catholic hero, Alick: "Gold will get a Dispensation, uncle," said Alick, "gold, the bright gold will do it – priest or bishop can't stand that by no manner o' means".[28] Significantly, his manner of speech, although there is no indication that he was meant to be anything more than the uneducated son of a prosperous farmer, is more refined than that of the priest who dismissed a dispensation as 'clane out o'rason', or that of the Bishop who, we are told, spoke 'fine English', and yet referred to a dispensation as "a great expince", and revealed his preoccupation with money, adding, '"those who expect the like favours from the Church must help support it"'.[29] Mrs Hall's bitter words about the priest and the bishop, and the pleasure with which the critic in *John Bull* received them may have had much to do with the political developments which had taken place since she wrote the *Sketches* in 1829. In that year a Bill was passed which enabled Roman Catholics to take their place in Parliament. 'Catholic Emancipation', as it was known, had come about only after years of agitation and the mobilisation of popular feeling by the Irish politician Daniel O'Connell. Irish priests were prominent in the movement for Catholic Emancipation, and were, with O'Connell, objects of fear and suspicion in English and Irish Protestant eyes.

Several tales in the second volume are of particular interest. One is the story of Murtagh Delaney, who is almost ruined by drink ('"the raale curse of the country"', in his fiancée's words) but who is saved by '"going to a good gentleman well known in these parts – one Mister George Carr – and writing his name in a book, promising not to touch a drop o'speerits"'.[30] The lesson is that drunkenness can be overcome with the help of those who care, but who are not family or friends. This is an early reference in Irish fiction to the temperance movement which had already been initiated by, among others, the Reverend George Carr, a family connection of Mrs Hall's, and was to be followed by the total abstinence movement which swept the country under the influence of the charismatic Father Mathew. It is a story suffused with an optimism that is lacking in most of the other sketches. There is genuine humour in the story 'Dermot O'Dwyer' whose rickety construction contrives to include several Irish anecdotes which had passed into the common store of story-telling and also a vigorous sermon from a priest exhorting his flock to search for missing money belonging to the local lord. The money is found and returned, demonstrating either the essential honesty of the peasant Irish, or their fear of the priest, or both. The sermon contains a pleasing little authorial jest – a private one between friends. The priest addresses a member of the congregation, thundering reproachfully '"Martin Doyle, is the horse gone lame, that ye never send a sod o' turf to my poor place, and yer own rick built up as high as the Hill o' Howth! Oh! Martin, Martin, yer a bitter sinner and so was yer father before ye!"'[31]

The joke here is that 'Martin Doyle' was the pen-name of the Reverend William Hickey, Rector of Mulrankin, a renowned agriculturist, who had set up a farm school in the parish of Bannow. His agricultural writings and his educational activities were so well thought of that he received the Gold Medal of the Royal Dublin Society in 1825, and he was, in every way, the very model of a progressive farmer. He was co-editor of the *Irish Farmers' and Gardeners' Magazine and Register of Rural Affairs*, which flourished from 1833 to 1840, and was well known for his writings, both in pamphlet and in article form. The Hickey family were close friends of Mr and Mrs Hall – the connection being through the Boyse family of Bannow House in Wexford, with whom Mrs Hall was friendly, as Samuel Boyse was also keenly interested in agricultural education, and received the Gold Medal of the Royal Dublin Society on the same occasion as did Hickey – and were among those invited to the Hall golden wedding celebrations in 1874. A daughter born to the Reverend and Mrs Hickey in 1839 was christened Anna Maria, presumably in a compliment to Mrs Anna Maria Hall, so it would seem that no offence was taken and that the families and friends laughed at their little private joke. Mrs Hall has another reference to the Reverend William Hickey in *Lights and Shadows*, when she says that the story 'The Last in the Lease' was told to her 'by a clergyman, who, under the name of "Martin Doyle" has published a variety of little works upon rural and domestic economy, the value of which, to the Irish cottager and farmer, is greater than pure gold'.[32] These works with titles such as *Commonsense for Common People*, *Irish Cottagers*, and *Hints for the Small Farmers of the County of Wexford* were indeed full of practical advice but one wonders how many cottagers actually bought them. Conscientious landlords, their stewards and agents and reasonably well-to-do farmers would have been the most likely buyers.

In the Preface to the second volume of *Lights and Shadows*, Mrs Hall states that the 'following "Sketches on Irish Highways" were written after a long residence in England' and her delight at being once more in her own dear Ireland was 'more than calmed by the misery which the poorer classes exhibit and which strike an English traveller's observation. My stories, therefore, have far more of "Shadows" than of "Lights"'.[33] She was, at this stage in 1834 and 1835, no longer a carefree young girl, but a married woman with her share of troubles. There had been the pregnancies which had ended in disappointments, the death of the only surviving infant, and the constant worry about her husband's career. Henry Colburn had appointed Mr Hall editor of the *New Monthly Magazine* in 1830 in place of Thomas Campbell (author of poetry which was much admired at the time and three of whose verses – 'Hohenlinden', 'Lord Ullin's Daughter' and 'Ye Mariners of England' – are still remembered with affection). Campbell was notoriously inefficient and S.C. Hall recalls in his memoirs that his editorial study was 'a mass of confusion; articles tendered, good or bad, were sometimes, after a weary search, found thrust behind a row of books in his study, and he was rarely known to give an immediate answer, yes or no'.[34] (Campbell's rejection of Miss Mitford's *Our Village* in 1819, while he was editor, probably owed more to lack of order than lack of editorial discernment.) The competent and conscientious Hall had, as Campbell's assistant, done his best to keep the magazine running smoothly, and his appointment as editor was a just reward for his efforts. Unfortunately, after a year, a great part of which he spent in clearing up the muddle and mess left behind by Campbell, Hall was removed from his editorial chair and his place was taken by Edward Lytton Bulwer. This fashionable novelist was brought in by Colburn, on the strength of his popular success, and his high social connections. Bulwer's editorship was not a success, however, and Colburn re-appointed Hall to the editorship in 1833, and he remained as editor until 1837, when he was displaced by Theodore Hook. Mrs Hall would have had good reason to feel insecure, and to worry about her husband's career and his future. It was natural so, for her to look at Ireland with new eyes, and to perceive what she may never have noticed before – widespread poverty and naked misery.

In any event, great changes had taken place in Ireland since Mrs Hall, then Anna Maria Fielding, had left in 1815. The boom years of agriculture during the Napoleonic Wars had given way to agricultural depression, and in spite of hopes raised by the Act of Union between Ireland, England and Scotland which had been passed in 1801, no industrialization had taken place in Ireland. Shortage of capital was a major factor here, and blame for this was laid by outside experts and economists of the day on the unsettled state of Ireland. In a long footnote to the sketch 'Beggars' Mrs Hall quotes from a pamphlet by 'an able and accomplished officer', Colonel Colebrooke, who sees the reluctance of foreigners to invest in Ireland as a direct consequence of the violence in the country.

'No capitalist', he says, 'will speculate in a country where the only return he can hope for his outlay, is to be murdered if he asks for it; no solvent agriculturist will settle in a place where his house is burnt down if he attempt to improve his land; no sober industrious man will take a farm where his life is in imminent peril, day and night, from the idle, drunken, turbulent tenants who were ejected from it. The very first step, then, we say, is to insure tranquillity and security, and then any plan of amelioration may be tried with a prospect of success.'[35]

Although it may well be true that the investors stayed away because of the violence endemic to Ireland, some of it at this time stemming from the Catholic majority's resentment at having to pay tithes for the support of Protestant parsons, it is something of a shock to find Mrs Hall quoting the rather intemperate words of the Colonel. Not only had she been at pains in her first collection of stories to emphasise how peaceful life in Ireland was, generally speaking, she had returned to the subject in 'The Last in the Lease' in *Lights and Shadows* and told a story within a story, for the express purpose of proving that Ireland was not a place of violence. Another long footnote here refers to a correspondence in *The Times*, shortly after the story had first appeared in the *New Monthly Magazine*. Her picture of a landlord's peaceful existence on his Irish estates was questioned by a reader, but Mrs Hall's reply (quoted in full in the footnote) stressed that there were peaceful areas in Ireland, where a gentleman might sleep safely in his bed and she gave the name and address of the person to whom she had originally referred. He was, she added, a resident landlord, amiable and excellent, and although 'politically opposed to the majority of his neighbours, respected and beloved by them'.[36]

One obvious result of the widespread poverty in Ireland, that 'utter hopeless, degrading poverty', was the large number of beggars who tormented every traveller and every passer-by, throngs of ragged, half-starved men, women and children who depended on charity to keep them alive. The three-month summer gap between the last of the old potato crop and the coming of the new was the worst time for unfortunates who were living at subsistence level. With no resources and no cash, there was nothing for labourers to do but go in search of work. This was more likely to be available in the eastern half of Ireland and was certainly so in England and Scotland, where extra hands were needed to help with the harvest. The labourers' families had to take to the roads and to a life of beggary. The problem was made worse, or at least more obvious, by the absence of a Poor Law, which did not come into operation in Ireland until 1838. Mrs Hall had written about beggars before in the *Sketches* and had spoken of them not only with some understanding, but with some humour, but of this mass of wretched beings she could only exclaim with pity and with horror. True, there were some individuals who were sprightly enough, because of their youth, and were able to make quips and sour jests and jostle others in their attempts to gain alms, but the overall impression was of starvation and disease. 'A pitiable sight', wrote Mrs Hall, 'a host of dirty starving creatures' in 'a moving mass of starvation and misery.'

Her genuine sympathy comes through when she writes of scenes such as these. There is no moralising, there is no lecturing the peasants on their foolish ways, there are no asides to the reader inviting complicity in condemnation of laziness or recklessness, just a simple human reaction of pity and rage at injustice. She addresses her English readers directly, saying plainly that it must be almost beyond their powers of imagination to understand the extent of Irish suffering. Those who live amid pleasant English scenes and abundant English luxuries 'may read of it in books – they may scrutinise it in pictures – but how completely do they fail to obtain even a remote idea of what it really is. The eye must see it – the ear must hear it – to conceive of its extent; or to appreciate its influence'. However, she must do her best to awaken English sympathy for her poor country and so she will continue to write about the poverty and misery she has seen. 'There is, in Ireland', she goes on

misery enough for gatherers of its records, without being confined to one subject. Misery is the refrain of Irish affairs. If we escape Scylla we fall in with Charybdis. The change of a Lord Lieutenant – the misrepresentations of an agitator – the cold denunciations of the opposite party, and the bitterness of each against the other – have comparatively little to do with the real state of Irish distress. There is positively nothing known, nothing imagined, of the utter, hopeless, degrading poverty endured by the peasants in the southern and wilder parts of Ireland.

As I write – while you read, there are hundreds of creatures gifted, unhappily gifted, with feeling and intelligence, yet having no prospect but starvation, no refuge but the grave![37]

Mrs Hall is careful to point out that her native region, the barony of Forth and Bargy, has escaped the ills that plague the rest of the country. She records that when she and her friends were making one of their tours they journeyed for more than sixteen miles through Wexford without meeting a single beggar or seeing signs of poverty and distress. But then, as she is never tired of repeating,

Lights and Shadows – *a melancholy book* 91

that district is home to many gentlemen who reside on their estates instead of leaving them in the hands of money-grabbing agents, and consequently, it is a place of 'peace, prosperity and contentment'. Alas, though, Bannow is not Ireland, and unless great changes are wrought it will remain an exception.

It is in her demand for changes that Mrs Hall shows a new sophistication and assurance in her writing. Humorous, tragic or melodramatic little stories are not now enough for her (although *Light and Shadows* has a fair share of these); she has to speak out more forcibly and clearly than before about the problems and woes besetting Ireland. A long passage lamenting the sorry state of the country is worth quoting in full as it pulsates with genuine feeling and honest, if bewildered indignation:

The ruined dwellings – the roofless cottages – the mismanaged farms – the improvident gentry – the trampled peasantry – in one sentence, the *ruined country*; the country over which foes triumph, and which "friends" betray – whose worst enemies are of its own progeny – whose sons may seek, and find, in every nation upon earth, except their own, prosperity and independence – whose daughters, conspicuous for wit, beauty and virtue, grace the courts of strangers, because the once gay and festive halls of Ireland are lone and desolate: the harps are hung upon the willows – the grass almost grows in the streets – the land is one of ruins, little prospers, even in its chief city.[38]

Mrs Hall's description of Ireland's malaise is as vivid as that of any of the foreign contemporaries who visited Ireland, and it is particularly affecting in that it is so personally felt and is the lament of one who knew the country in happier days. She has no easy solutions to offer, but concludes, 'It seems a mystery that centuries should pass and leave her [Ireland] more desolate and more depressed; yet, so it has been, and so – for aught I can see – so it will be for many a day to come'. This is a different Mrs Hall from the one who wrote the bouncing and extravert Irish sketches. She is still cheered by examples of individual Irish wit, kindness, hospitality and generosity, but the overall picture is so grim that she is near despair. Outside help must come – from England, whose government had a moral responsibility to the Irish nation – but that alone would not solve all the problems. The Irish people themselves must play their part and it was the landlords who were to give the lead. They must not only reside on their estates, they must administer them properly, taking advantage of all the new methods of agriculture which were working so well elsewhere, and they must care for their tenants in every sense, moral as well as physical.

Mrs Hall dedicated *Lights and Shadows* to Mrs Grogan Morgan of Johnstown Castle, County Wexford, 'not so much in testimony of private friendship' as in recognition of the good example she and her husband were setting other landowners. Mrs Hall prayed that they might find many imitators among landed gentry and that the tenantry might recognise the value of the Grogan Morgans' 'unceasing labours for their moral and social improvement'.[39] It was not only the landlords, so, who must work in Ireland, the tenants too must do their share. It would not be easy for them, but with enlightened guidance from their betters they could learn to overcome the difficult circumstances of their lives, circumstances which were, in some cases, of their own making. If they listened to wise words of instruction they could overcome their native faults and life would be immeasurably better. Mrs Hall was now more than ever aware of her duty as a teacher. English people had been addressed directly in *Lights and Shadows*, in plain language, and had been blamed for their neglect of Ireland; now it was time to address the Irish people. She was aware that the most important relationship in the country was that between the landlord and his tenants, and that was where she must expend her talents and energies as a teacher. Her only Irish novel, *The Whiteboy* is concerned almost exclusively with this relationship and in it the greater responsibility lies with the landlord. *The Whiteboy*, however, was not to appear until 1845, and before then Mrs Hall addressed the tenantry, pointing out to them how they could play their part in the regeneration of Irish life. Her advice appeared in a series of stories published in 1839–1840 as *Stories of the Irish Peasantry*, supposedly for the edification of the poor men and women, who in the words of the countryman she quoted in *Lights and Shadows* 'have a wedding, and little to begin with – a power of children, and little to give them – rack-rent for the bit of land, turned out bag and baggage for that or the tithe – beggary – starvation – sickness – death! That's a poor Irishman's calendar!'.[40]

Lights and Shadows marks a change in Mrs Hall's perception of Ireland, and is, in spite of some humorous and light-hearted sketches, a melancholy book. This was not what readers wanted from Mrs Hall – they would prefer to read about blundering Irishmen and slovenly Irishwomen to whom they could feel superior, and to whom they owed no responsibility. Significantly, when some of the contents of *Lights and Shadows* were reprinted, it was the humorous sketches illustrating faults in the Irish character that were chosen (e.g. 'It's My Luck' and 'Moyna Brady'), not the more sombre ones with their descriptions of a 'ruined land', and their imputation of English neglect and misgovernment.

SIX

Stories of the Irish Peasantry – Correcting the 'evil habits of poor Pat'

In her introduction to the Fifth Edition of *Sketches of Irish Character*, published in London in 1854, Mrs Hall recalled what her object had been when she first wrote her sketches almost a century before. It was, simply, to make her native seaside village of Bannow 'favourably known to the English', and, she added, 'I did not then think that I might be useful to my country by endeavouring to correct the failings of its affectionate and generous peasantry'.[1] In those early sketches and in several others contained in the collection *Lights and Shadows of Irish Life*, published in 1838, she highlighted Irish failings but it was not until 1839 that she wrote a series of stories in an attempt to correct them. The tales 'Stories of the Irish Peasantry' appeared throughout that year, and part of 1840, in *Chambers's Journal*, the periodical which had been founded in 1832 by Robert Chambers of the Edinburgh publishing firm W. & R. Chambers. The *Journal* was a very popular publication, catering to the contemporary appetite for information on art, literature and science and having a high moral tone. Mrs Hall's stories appeared alongside articles on subjects as diverse as simple economics, antiquarianism, poultry rearing, modern inventions, and household hints. There were also book reviews, poems, legends, short biographies and philosophical essays on the nature of happiness and the pleasures of leading a simple life. William Chambers in his autobiography, *Story of a Long and Busy Life*, published in 1882, recalls how he felt honoured by his acquaintanceship with 'Mrs Anna Maria Hall, the wife of Mr S.C. Hall'. Chambers often visited the Halls in their 'pretty little villa' in London and found Mrs Hall to be 'essentially Irish in her vivacity and geniality of disposition'. Of her work he commented: 'Her *Stories of the Irish Peasantry*, each with a distinct moral purpose ... were much appreciated by the readers of *Chambers's Journal* of which she was always an acceptable contributor'.[2] There were sixteen stories in all and almost every one dealt with a distinctively Irish failing – recklessness, improvidence, violence, drunkenness, etc. The last in the series, called 'Debt and Danger, (No. 2)' appeared in May 1840, and was followed by an editorial announcement which read 'The preceding tale concludes the series of *Stories of the Irish Peasantry* which Mrs Hall undertook to write for our pages with the patriotic view of improving the morals and general economy of her fellow countrymen, as well as amusing and perhaps instructing readers among the humbler classes in Britain'.[3] A collection of these stories was 'now in the press' and would 'immediately appear as a cheap volume in the "People's Editions"' (Chambers's own series).

The 'instruction of the humbler classes' was a familiar early nineteenth-century preoccupation in Ireland, England, Scotland and Wales. There was some resistance to the idea of working classes reading at all, because of the danger that they might be exposed to subversive literature such as that supplied by Thomas Paine and William Cobbett. If, however, the lower classes were going to read, then suitable literature should be provided for them. The tracts published by the *Religious Tract Society*, many of them written by Hannah More (Cobbett's 'prime old prelate in petticoats') were not enough; something more secular was required. Publications from the *Society for the Diffusion of Useful Knowledge* began to appear in 1827 and gave mechanical and scientific information in a simple form. They appeared twice a month, and were expensive at sixpence each. The *Edinburgh Review* had already stressed the importance of literature for the lower classes, saying, 'We try to impress upon the wealthier orders of the community the duty of promoting among their dependents and neighbours the circulation and perusal of such really useful publications', and concluded by recommending the labouring classes 'to the more particular attention of literary men who may well devote somewhat more of their time to instructing them in their duties, never forgetting their rights'.[4] In this context 'labouring classes' meant peasants, and landowners who took their feudal duties seriously circulated among their tenantry little textbooks, tracts and pamphlets designed to improve their physical as well as their mental condition. These were manuals of domestic economy, hygiene and good husbandry, set in fictional form to catch the popular attention, fed as it was on chapbooks and shortened forms of Gothic romances.

The best-known of these textbooks for tenants was *The Cottagers of Glenburnie* (1808) by Elizabeth Hamilton, an author who had already achieved some success with her novel *Letters of a Hindoo Rajah* (1796). The Scottish cottagers of the title are a slovenly lot whose way of life is transformed by an energetic middle-aged, middle-class lady who comes to live among them and who, after initial difficulties, convinces them that her ways are best. It is a highly moral tale, wherein the wicked are punished and the good are rewarded, but the practical advantages of careful housekeeping, cleanliness, hard work and determination are stressed with almost equal force. The whole assumption of the book is that poor and ignorant people are incapable of bettering themselves by their own efforts, but must depend on those of a higher social class to help them, and the *Edinburgh Review* commented that 'some good [might] be done by the circulation of this work among the lower classes of society'.[5] Almost forty years later, Samuel Ferguson, writing about William Carleton's *Parra Sastha*, in the *Dublin University Magazine* in 1845 referred approvingly to the effect which *The Cottagers of Glenburnie*, 'that lovely little Scottish tale', had had upon 'the upper classes of that country, by directing [their] attention to efforts for the preservation of cleanliness and decency among the populace'. (He doubted, however, whether it had ever had 'much influence in reforming the sluttish habits of the lower classes'.)[6] Mrs Beatrice Grant's *The History of an Irish Family*, published in Edinburgh in 1822, was in the same fictional form as *The Cottagers*, but lacked the immediacy and freshness of the earlier work. Practical advice is considerably outweighed by moral precepts and the prose style is leaden but the work is of interest if only for the author's reference to 'Lady L's book of instruction, which was gifted to every family on my Lord's property', and was a 'simple but most beneficial work on Cottage Education'.[7]

The term 'Cottages' and 'Cottagers' carried their own overtones of landlord/tenant relationships and were used by Irish as well as English and Scottish writers. Mary Leadbeater, a member of a well-known and respected Quaker family from County Carlow, wrote a series of little books designed to educate and improve the peasantry under the title of *Cottage Dialogues among the Irish Peasantry*. The first of these, with a Preface by Maria Edgeworth, was published in 1811 and was greeted by the *Eclectic Review* as being not only an accurate description of Irish society, but likely to be of great value to the Irish peasantry as it 'taught a great number of important moral and prudential lessons in a form peculiarly attractive and striking'.[8] No critical attention was paid to any similar didactic fiction for Irish peasants until the Reverend William Hickey, the Church of Ireland clergyman from County Wexford and friend of Mr and Mrs Hall, wrote his *Irish Cottagers* under the pseudonym 'Martin Doyle' in 1830. This collection of cautionary tales has much in common with Mrs Hamilton's

Cottagers of Glenburnie in that the practical advice is sound and sensible, and the prose style, including the use of peasant dialect, is sufficiently attractive for the stories to make easy reading. The *Literary Gazette*, reviewing Martin Doyle's *Irish Cottagers*, commented that it supplied 'a vacancy in our literature too much neglected. We cultivate and uphold a taste for reading in the lower classes, but we do not sufficiently attend to the necessity there is also for supplying wholesome and nutritive food to the appetite thus created'. This book, however, was admirably suited to that purpose, the pages being 'full of useful instruction and practical example'.[9] The Irish *Christian Examiner*, likening Martin Doyle to Cobbett in the vigour of his prose, but without 'his impiety or disloyalty', saw the books as ideal for distribution by clergymen among their parishioners, as they were 'amusing while [being] instructive'.[10] The *Athenaeum* welcomed *Irish Cottagers* as 'a didactic tale for the amusement and instruction of Irish Cottagers, and altogether well adapted to convey improving sentiments to a class greatly in need of them', and the critic praised the simplicity of the stories which recommended 'cleanliness, sobriety, industry, and the other virtues, the lack of which is the cause of the people's misery'.[11]

Mrs Hall's *Stories of the Irish Peasantry* are very much in this tradition of instruction, but they lack the subtlety and skill shown by Elizabeth Hamilton, Mary Leadbeater and Martin Doyle. Although explicitly addressed to the Irish peasant, some of these stories are comments on Irish failings to a third party, and in others the author herself is present, as she was in the earlier sketches, 'The Bannow Boatman', 'Independence', 'We'll See About It', and 'It's My Luck'. This is a device which makes the tale lack conviction as a piece of fiction when it is often repeated, and causes the moral to be pointed in a mechanical way. A truly effective teaching story does not depend on a narrator who is present and pointing out to a pupil where the mistakes have been made that precipitated the tragedy. Other stories in the collection are very much in the tract tradition and end with a warning that sums up the story, for example, 'Keep clear of DEBT, and ye'll keep clear of DANGER', and 'ONLY A DROP is a temptation fatal – if unresisted'. (It is only fitting that *Stories of the Irish Peasantry* be included with a bundle of miscellaneous tracts in the British Library.) 'It's Only a Drop' is, of course, a cautionary tale about Irish peasant drunkenness. It's a predictable story, with echoes of 'The Groves of Blarney' – the young heroine refuses to marry her admirer until he swears never to touch even a drop of whiskey – but the young man in this case is merely frightened off whiskey by a tale told to him by the heroine. It is the story of a 'well rared' woman known as 'Lady' Stacy because of her genteel manners and mode of speech. Her childhood had been blighted by her father's addiction to drink and her only child died as a result of her husband's negligence while drunk. It was a truly melodramatic story told with energy and passion and much narrative skill. 'Lady' Stacy described her father's carousing (while his wife lay on her deathbed, and the bailiffs were beating on the door) in the parlour where his drinking companions had '"raised a ten-gallon cask of whisky on the table ... and astride of it sat her father, flourishing the huge pewter funnel in one hand, and the black jack streaming with whisky in the other; and amid the fumes of hot punch that flowed over the room, and the cries and oaths of the fighting, drunken company, his voice was heard swearing 'he had lived like a king, and WOULD die like a king'". He did not die like a king, of course, but '"died smothered in a ditch, where he fell"'.[12]

The same vigorous language is used in the description of the infant's horrible death, burnt beyond recognition in a fire caused when its drunken father dropped a candle into the straw around its cradle. The mother, returning from an errand of mercy, saw '"a white light fog"' coming out of the '"dark mass"' of the cabin. She '"darted forward as a wild bird flies to its nest when it hears the scream of the hawk in the heavens"' but was too late. She fled to the doctor with '"what had been her child,"' but "there was no breath, either cold or hot, coming from its lips then" and she '"wandered the whole night long in the woods with that burden at her heart"'. Both tragedies were caused by whiskey: '"the curse of Ireland – a bitterer, blacker, deeper curse than ever was put on it by foreign power or hard-made laws"'. This story within a story (one of Mrs Hall's favourite narrative devices) could have stood on its own and been effective but it loses its impact by being placed within the larger flimsy framework and by having the moral spelled out and underlined.

Dirty Irish ways do not merit a complete story of their own but are referred to several times in the collection. In 'Time Enough', a story which illustrates Irish laziness and fondness for procrastination, the slatternly Judy Radford, mother of 'seven lazy, dirty, healthy children' lives with her brood in a cottage, that could hardly be reached 'for the slough and abominations that surrounded it'. The narrator rebukes her: '"Such a heap of impurity must be unhealthy"', but Judy points out that '"the young ducks would be lost without it"'.[13] In 'Going to Service', a story that deprecates false Irish pride, a young girl, Mary Mulvany, spends some time in the home of a distant relative who has no idea of how to manage a household and who is generally indifferent to dirt and disorder. Mrs Hall gives an amusing description of the ménage:

> The kitchen was a scene of most desperate confusion; instead of the noggins and jugs being hung in regular lines along the dresser, they were laid down when done with on the floor, 'that the cat, the craythur, might finish the sup of milk', or 'the chickens pick the last of the stirabout', or 'Rover, the baste, lick the end of it'. There they remained until they were wanted, when all was perplexity to get them ready. The dirt was never disturbed from the corners of their parlour, or from behind the tables and the chairs; consequently every breath of air that entered the room set it whirling over 'the greenest spot' that had received the *promise of a sweeping*.[14]

Mary had been properly brought up and was horrified by what she saw, but worse was to come. She 'discovered in the morning, while commencing her breakfast, that the milk had not been properly strained before it was set for cream to make butter; consequently the cow hairs stuck round that compound like a *cheveau-de-frise*'.

A vivid picture, and one of particular interest. The cleanliness or otherwise of butter in a peasant society was not only a matter of hygiene; it was of great social importance – carelessness in butter-making was the mark of a slattern. In Elizabeth Hamilton's *The Cottagers of Glenburnie* the butter made by the sluttish Mrs MacClarty so revolted her fastidious visitor, Mrs Mason, that she could not eat it because 'her disgust was augmented by the sight of so many hairs, which bristled up upon the surface of the butter, when spread'. William Carleton, that observer of peasant life from his place within it, also noted the importance of good butter-making; in his sketch 'The Station' in *Traits and Stories of the Irish Peasantry*, published in 1830, the woman of the house, Katty, 'bore so indifferent character in the country for cleanliness, that very few would undertake to eat her butter. Indeed, she was called *Katty Sallagh* [Dirty Katty] on this account'.[15] In Mrs Hall's story young Mary is accustomed to habits of cleanliness because she had been taught them by an Englishwoman living in Ireland, while Carleton's people have a natural aversion to dirt. These contrasting pictures are yet another example of the different way in which the two authors describe peasant attitudes.

Travellers in Ireland in the early nineteenth century almost invariably commented on the numerous children who were to be seen everywhere. This was a fact, not merely an observation. There was a rapid population growth in Ireland between 1780 and 1830, which historians now attribute to a combination of factors – low death rate, high marriage rates and high fertility in marriage. The widespread use of the potato made early marriage possible because enough potatoes could be grown in a small plot to feed a large family, and a pig, and a diet of potatoes was a healthy one. Mrs Hall deprecated the fact of early marriages and the production of numerous children because she saw it as leading to family misery – a not unreasonable viewpoint. In earlier works she had commented on what looked to her like recklessness and lack of forethought, and in 'Too Early Wed', the first tale in *Stories of the Irish Peasantry*, she warns of the sad consequences of an early marriage. Sandy is a twenty-year-old servant, who is borrowing money from his lady employer in order to wed his sweetheart Lucy. The opening dialogue between Sandy and his employer conveys Irish peasant attitudes to marriage. Reason collides with romance as his employer asks Sandy what provision he has made for matrimony:

"Provision is it, my lady?" answered Sandy, with another turn of his hat; "we've lots of love, misthress dear; it'll hould out till the grave shuts over us, I'll go bail for that."

"But, Sandy, you can't live on love?"

"It's cruel poor living without it – that I know ma'am" he replied right steadily.

"But there will be two to feed instead of one at your father's, for Lucy cannot continue at the lodge."

"Nor doesn't want, ma'am – I've built her a cabin off the corner of my father's three acres, and there's a few sticks in it already. She's no great eater, and the pratees are cheap enough, thank God!"

"But by and by, you will have more than two to feed."

"Please God," was Sandy's quiet reply.[16]

In that short passage of dialogue, Mrs Hall sums up contemporary Irish peasant marriage beliefs. Love is important, living is cheap, and children are a blessing from God. Sandy's employer tries further to dissuade him, pointing out that Lucy is not yet eighteen, "'too young to take the heavy cares of peasant life upon her'". She begs him to wait for a few years so that they can both save some money that will "'furnish a cabin comfortably, and a short purse to defray the first expenses'". He is not to be dissuaded, however, not even when his employer (who is the narrator of the story) appeals to his genuine love for Lucy and begs him not to condemn her to a bleak future. "'Marrying so young, old age will come upon her prematurely. Her eyes will grow dim, and her hair turn grey before her time; her bodily strength will fail,'" and she will not be able to augment their small income by work of her own for '"what woman can knit, or spin, or sew for hire, with a tribe of little half-starved children round her feet?"'

All argument is useless, and Sandy determines to marry Lucy, who is equally firm in her intention to marry at once. The narrator muses on their folly in 'acting from impulse, rather than reason' and states firmly that 'these early marriages are sources of the great evils of Ireland, and can never be prevented, as long as the peasantry have no ambition to elevate themselves in the scale of society, by means of better clothes and better dwellings than they generally possess'.[17] Mrs Hall cannot resist the temptation to moralise, instead of letting the story speak for itself. Lucy's marriage ends in tragedy as had been predicted. Her husband became ill, a baby arrived, and then another, and the family sank so low that Sandy went to England to find work. Before he could send home any money, however, Lucy was driven by near-starvation to beg on the roads, and finally died of cold and hunger, at the very age, commented the narrator, 'when, if she had followed our advice, she might have married in sure anticipation of happiness, and with a reasonable prospect of prosperity'.

If Irish people were reckless and lacking in foresight they were also, particularly the peasants, deeply conservative. No matter that this is an attribute of peasant society world-wide, Mrs Hall chose to see it as peculiarly Irish. 'Sure it Was Always So' illustrates this conservatism, or lack of initiative. The women have to go a mile for fresh water because the well near at hand caved in many years before. It would be an easy matter to clear it out, but the women had become used to the inconvenience. Furthermore, they believed that there had been a curse laid on it in earlier times, so superstition allied itself with inanition. The narrator used a pick-axe to clear out the old well, and no harm came to her. By her example and her argument she convinced the women that there was nothing to fear and that life would be much easier with a supply of pure fresh water in the village: 'The well was done, a comfort bestowed, and a superstition overcome, at a very small expense of time and trouble'.[18]

Mrs Hall's friends, the Grogan Morgans, owners of Johnstown Castle, who were celebrated by her in the sketch 'The Landlord at Home' had difficulty in making their tenants adapt to new ways – 'Paddy likes to go on in the old way' – and the author tells the story of how 'the sapient occupier of a cottage with a floor composed of strong lime cement', decided he 'would rather thrash his corn in that room than in the shed; and as the ceiling interfered with the action of the flail, and Paddy could not practice ... he *dug a deep hole in the floor*, and in this hole was he thrashing away right merrily, to the tune of "The Rakes of Mallow" when his fair young landlady entered!'[19] "Easy Jack" Cummins was asked by his landlord in the sketch 'Family Union' to combine with his neighbours in draining a three-acre field which was divided amongst them, but although he was assured the improvement would not result in a higher rent, that indeed he would be forgiven a year's rent, he never did so. His excuse when asked about it later on, was: '"My father left it the way it was in his time, and my mother found it mighty convanient for rearing young ducks and blaiching flax"',[20] so it was left in its undrained state.

Mrs Hall's assessment of peasant conservatism was accurate, and she described it amusingly enough but it should be taken in conjunction with William Carleton's writings on the same subject, especially his short novel *Parra Sastha*. This avowedly didactic work, published in 1845, is the story of a farming family who, although substantial landowners, were so lazy, dirty and obscurantist that they lived in squalor and discomfort and allowed their land to lose heart. It sounds unpromising, a mere moral tale of a man redeemed by the good woman whom he marries, yet it is a work full of interest for its humour, and its keen observation of peasant character. Paddy, his parents and his sisters, the whole Go-Easy family, although presented as caricatures, come to vivid life in their dirty, stinking cottage surrounded by neglected farm buildings and badly cultivated fields. The elder Go-Easies sit smoking happily by the fire while the girls giggle and titter away in the background, oblivious to the dirt. Everything is exaggerated yet it is believable, probably because of the dialogue which reveals the peasant mind in all its blind ignorance. It is in this that the true savagery of Carleton's attack lies because it subverts all the romantic notions that might have been held about native peasants' nobility. Here are real people talking, bitter, mean-minded and begrudging as they jeer at a neighbour's improvements. The speaker can barely talk for laughing but between his convulsions he manages to tell the others that their neighbour, Denny Delap, has '"got an *Iron Plough*, no less; wouldn't put up like his neighbours wid' a wooden one. There's grandeur for you"'. His listeners are amused and scornful and everyone agrees that Denny has become too much of an upstart. He is mocked because of his appearance, '"plain little Dinny that has the cast in his eye"', but it is clear that it is his attempt to improve his farming methods that rankles, and it is recollected that "the same dirty little scrub" had once laughed at the eldest Go-Easy for '"harrowin' part o' [the] oats wid' a thorn bush an' a stone on it"'. The company concludes that '"he'll reap sorrow for the same plough, as far as makin' him an' it the standin' joke o' the whole counthry goes"' and they all sit back and smoke to their hearts' content. The author comments:

> Such was the reception which anything connected with domestic or agricultural improvement was certain to receive from the Go-Easy family, not one of whom was capable of comprehending an enlightened principle on any subject whatsoever, especially if it happened to go beyond the old stated landmarks of their own ignorance, or transgress the dark and limited range of their experience.[21]

Paddy, however, for all his faults and deficiencies, is good-natured and malleable. He marries Nancy after a long courtship, and she is the exact opposite of Paddy; neat, orderly, industrious and with progressive views on agriculture as well as on domestic matters. Her reception in the filthy house, where even her spirit fails for a time (she has to break a pane of glass to let some fresh air into the foetid bedroom), and her gradual reformation of Paddy and his sisters, is described with such skill and such humour that it sweeps the sketch along and lifts it out of the mould of the tract. Nancy represents the Irish peasant at her best and Paddy shows that even the laziest peasant can be reformed if the task is undertaken with diplomacy – and with affection. Paddy is never made to feel inferior and Nancy never commands or bullies but gets her way by working on his good qualities.

Carleton is considerably more condemnatory of peasant behaviour than was Mrs Hall, and more extreme in his language, but the essential difference between the two literary teachers (for such they were) was that he believed reform of society could come from within while she saw it coming from without. It is a peasant girl who reforms Carleton's Paddy, a girl from his own class, while Mrs Hall's archetypal Paddy must be taught from without. Her Paddy needs either a resident landlord or a neighbour of English descent to show him how to improve and mend his ways. English superiority had been cited by Mrs Hall in her previous works, the *Sketches* and *Lights and Shadows*, and in *Stories of the Irish Peasantry* it is adverted to again. Mary Cassidy, the young girl in 'Going to Service' received good advice from a neighbour who had been trained in proper housekeeping by her English mistress; Paul Kinsala in 'The Landlord Abroad' paid his rent on time 'like an English tenant', Pierce Scanlan, whose boasts are the subject of 'It's Only A Bit Of A Stretch', grumbles when his aunt 'a good, quiet English soul' tries to make him listen to reason, and in 'Sure It was Always So' it is a visitor from England who clears out the disused well. The good landlords who live on their estates and who help their tenants to better themselves, are, it is implied, following English feudal practice.

In all her works Mrs Hall praised Irish hospitality and generosity but in *Stories of the Irish Peasantry* she warned against the dangers of imprudent generosity, especially when it sprang from lack of forethought. In the sketch 'It's Only the Bite and the Sup', for instance, she told the story of a peasant woman who had 'a family of five children, a blind grandmother and a lame husband' and yet had sheltered a piper, his wife and son for a month and fed them from her own small store without considering the consequences. Typically, the narrator comments: 'The poor cottager must not persuade himself, that if he gives his own and his children's food to the poor traveller, he wrongs *none but his own*. Society is so constituted that we cannot wrong *only ourselves*; those who give all give none. When Mary Flanagan supported the piper, his wife and child for a month, having barely enough to feed her own family until the potatoes came in she *created beggary*'.[22] In the same sketch there is the story of Jenny Jeffers, a comfortably-off young girl who married a ne'er-do-well, and saw him squander all her money on drink and on boundless hospitality. The author commented: 'under the generosity of such profusion, there frequently runs an undercurrent of *love of praise*, which stimulates persons not high-minded enough for liberality to a reckless extravagance'.[23]

As a self-confessed lover of Ireland and the Irish people, however, Mrs Hall was eager to emphasise Irish virtues, especially those she appreciated among the peasantry and she had a particular admiration for the way in which the poor, 'the warm, kind-hearted people, who have no ungenerous faults', helped one another. Her tragic peasant characters, such as Lucy in 'Too Early Wed' speak of neighbours sharing the little they had 'even to a handful of meal, or a stone of potatoes', and in 'The Follower of the Family' strangers looked after the nurse Margaret Sheil and her foster child, little Evelyn, when they fell ill with fever on their weary way to Dublin in search of a surgeon to cure the child's blindness. Mrs Hall described the way in which fever sufferers were tended in Ireland:

The peasantry never totally desert each other; they dare not, of course, bring the infected party to their houses; but before the next morning dawned, this good Samaritan, had, with the assistance of a neighbour, erected a sort of shed over the sufferers so as to protect them from the inclemency or heat of the weather, and placed a comfortable quantity of dried heath beneath them.[24]

Milk was 'pushed towards them with a long wattle every day' and some simple food when the fever broke. Carleton describes a similar scene in *The Poor Scholar* when Jemmy McEvoy is stricken by fever and looked after by strangers.

Mrs Hall also lauded the purity of the women and the affectionate family feeling of the men, but her greatest admiration seems to be reserved for the fidelity shown by a servant to a master. This was one of the main themes in the story of *Harry O'Reardon*, the short novel published in Philadelphia in 1836 and later included in *Lights and Shadows of Irish Life*, and she returned to the subject in the long short story, 'The Follower of the Family' in the *Stories of the Irish Peasantry*. In this story, the maid servant, Margaret Sheil, looked after three generations of the O'Dwyer family, nursing and sustaining them through poverty, illness, death, exile and blindness, seeking no reward but the affection of her charges and her own consciousness of doing what was right. She was even willing to give up for adoption to a rich and kindly lady, the little girl, the last of the O'Dwyers, whom she had reared. It was for the child's own sake, and Margaret would not stand in her way: '"I'm deeply grateful she should keep with those who can put her in her own station; and I'll be no burden on them or her"'.[25] In another story, 'Debt and Danger', an old retainer who tried in vain to check his young master's extravagance remains faithful to him throughout all his tribulations, even going so far as to hide him safely when the bailiffs are seeking him.

Mrs Hall is at something of a loss when it comes to what she sees as the natural quick-wittedness of the Irish. She can not help admiring it but she is exasperated by the way in which logical arguments and reasonable suggestions are turned aside by clever words and jests. She says ruefully that

> Paddy is so full of humour, real genuine humour, that he will lean his back against the door-post ... put one foot over the other, take his *dudeen* out of his mouth, fold his arms across his ample chest, and beguile you from the intention of giving him a good lecture both on the management and mismanagement of his farm, until you wish him good evening, enjoying the remembrance of the raciness and humour of his stories, and the mirthfulness that shakes his rags with laughter. It is not till after you sit down to your reading table that you think how completely you were beguiled of your wisdom![26]

The stories in this collection have a historic or sociological interest – they throw some light on a certain section of society at a certain time, but it is difficult to find much literary value in them, they are dull, lacking in dramatic tension and overloaded with advice and admonitions. The prose is ornate and lacking in the simplicity of the earlier sketches, especially in the descriptions of landscape. For example, a mountain that she once styled 'handsome' and 'picturesque' in an earlier story, becomes 'a great magician', 'noble' and 'time-honoured' and where before it sheltered 'squatters' who built cottages and tilled the ground on its slopes, it is now inhabited by 'hardy mountaineers; fine specimens of the animal creation', who are 'as picturesque as brigands on the finest Italian crags that were ever painted'. The 'pair of green islands ... called the "Keerooes", where in summer, a few starved sheep or one or two goats, wander over about an acre of moss and weeds', which were the scene of a shipwreck in 'Kelly the Piper' in the *Sketches* have been transmuted into 'emeralds in ... crystal waters, which chafe and fret against the dark rocks' in 'The Landlord at Home'.[27]

Very few of the peasant characters created in the *Stories* by Mrs Hall have any semblance of reality, especially in their speech which is either in the stage-Irish mould, or is a form of sermon rebuking others for their faults and shortcomings. One of Mrs Hall's earlier strengths was the manner in which she reproduced Irish peasant speech in all its flamboyance and hyperbole and yet managed to make it seem natural (if stereo-typical) but in these stories her skill seems to have quite deserted her. This is most obvious in the two companion pieces 'The Landlord Abroad' and 'The Landlord At Home.' They are addressed to the gentry, not to the tenants, of course, but it is the peasants who do all the talking, and their speech is totally unconvincing. It is not only the reeking servility of the sentiments that jars on the reader, but the form in which it is expressed. In the story of the good landlord – the one at home – there is an authorial eulogy of Mrs Hall's own good friends, the Grogan Morgans of Johnstown Castle in County Wexford, followed by comments made by their supposed tenants, Anty and Abel. These two have quarrelled over the improvements made and proposed by the landlord, but Abel comes round to Anty's more progressive viewpoint, and praises the owners of Johnstown Castle:

"I'd rather", said the lover ..."I'd rather that castle was on the top of the Mountain of Forth, as an example to the country, than sunk down in a valley."

"Sure, it is as well where it is; has as fine a *moral* influence, the people, I mean, that's in it."...

"MORAL INFLUENCE!" repeated Abel; "I daresay that's the right sort of thing; but I'd have the advantage of every sort of influence given to such people. The castle, I tell you, should be on the top of its mountain in its glory."

"The glory of their good deeds will go higher than that," said Anty.

"I know – to Heaven!" replied the young man. *"But for that I'd have him a lord on earth."*

"They'll be saints in heaven", said the girl.[28]

Many of the *Stories* are based on the same plots as those used in the *Sketches* and a stale air has taken the place of the earlier freshness. This may account for the fact that there were no reviews of the work in English or Scottish periodicals. The only direct comment was made in the *Dublin University Magazine* in 1839, taking note of the stories as they appeared in *Chambers's Journal* that year. Mrs Hall has, according to the reviewer, 'a power of brief and sketchy tales which is inimitable' and 'her anxious wish seems to be to do good and to be as useful as her means enable her'. As, however, she has neither great wealth nor political power, she has put her pen at the service of her countrymen and women, and has tried to correct 'all the bad habits that mar their domestic comfort'. Each 'tale illustrates some peculiarity or evil habit of poor Pat' but although the reader may be amused by some of Pat's ways 'it is the smile of goodwill which a kind person bestows on failings which he regards with as much love as pity'. The reviewer commends the stories to each landlord, pointing out that he 'may buy one of these for *three halfpence*, to circulate among the poor peasantry of his neighbourhood'.[29] At this distance, there is no way of knowing how many landlords took the DUM's advice and distributed copies of *Chambers's Journal* to their tenants, and even less chance of finding out tenants' reaction to these printed homilies. Even at the time no one could tell if Mrs Hall's teaching would have any effect. This was the view of *Chambers's Journal's* editor, expressed in a notice in the periodical promising the publication of the stories in book form. It is impossible to say, he concludes, how well Mrs Hall may have succeeded in her 'generous design' of correcting Irish faults, but, he adds, 'from the extensive circulation of the stories in Ireland, and the manner in which they have been received in this country, we have little doubt that the result has been beneficial'[30]. Sales were enough to warrant a second edition which was published in 1850, and dedicated by Mrs Hall to Irish landlords and their tenantry in the hope that she 'might benefit both by showing that the interests of both are mutual and inseparable'. Her 'heartfelt desire' was to 'serve her humbler countrymen and countrywomen' and she trusted that 'the cheap publication ... submitted to them might be placed by the higher class within reach of the lower'.[31] *Stories of the Irish Peasantry* did not enhance Mrs Hall's literary reputation but her popularity with readers of *Chambers's Journal* was not diminished, because she continued to write for the periodical for many years, and collections of her stories were published by the Chambers brothers in their 'Miniature Library of Fiction' series (1858). In Ireland, however, a sharp criticism on a personal level, made in the nationalist periodical, the *Citizen*, in 1841, gives a hint that readers may have become weary of being lectured. In a review of a new edition of Mary Leadbeater's *Cottage Dialogues Among the Peasantry* reference was made to Mrs Hall as one of those who wrote books about the peasantry of Ireland. 'Then cometh', the critic says, 'another tribe of foreigners, quacks who declare themselves upon a "mission" to cure the faults that so divide Ireland from the civilised portion of the world; and Mrs S.C. Hall writeth "Stories of Ireland" as she calls them, by way of giving good advice to "her countrymen", quite confident of persuading them to imitate herself, and call the country West Britain as fast as possible.'[32] The truth is that Mrs Hall's books were never as popular in Ireland as they were in England; we have Mr Hall's word for that. In his memoirs he puts this down to the fact that she 'belonged to no party and took no sides, being neither Orange nor Green',[33] the implication being, of course, that unlike Irish readers, those in England were above such political and religious distinctions and could appreciate the talent of the story-teller, herself impartial and free of all shades of bigotry. Mrs Hall's tolerance of political and religious differences was emphasised in the *Dublin University Magazine's* assessment of her career which appeared in their series 'Our Portrait Gallery', in August 1840. She was the first woman to be so featured and 'a lady', said the writer (W.A. Butler), 'in whose commendation we believe all men of all parties can cordially unite'. Her works were free of all 'party feeling' – not that that of itself was 'necessarily a merit – for, no doubt political and religious truths may be most forcibly and usefully inculcated through the medium of fiction' – and in Ireland that was a great blessing for there were far too many partisan novels on sale. 'In the case of Mrs Hall', adds the writer, 'it would be impossible for the most irritable of sectarians to find any sentiment, political or polemical, too sufficiently violent to find fault with.' He attributes this tolerance on her part to the fact that she left Ireland while young, and so her pen never 'drew a sketch but was perfectly free from either of the two tints, orange or green, which seem almost natural to an Irish picture'.[34]

Although *Stories of the Irish Peasantry* did not make much impact on the literary world, the same cannot be said of Mrs Hall's next book, *Marian, or a Young Maid's Fortunes*, published in 1840. It was a critical and popular success and reinforced her position as an interpreter of Irish life with its sympathetic, and idealised portrait of an Irish servant. Critics singled out this feature of the novel for special praise and once more commended Mrs Hall for her 'Irishness'. *Marian* is an interesting book, for another reason. It gives a very detailed picture of life in an English boarding school for girls, that may possibly have been drawn from personal experience. We know nothing of Mrs Hall's life in England from the time she left Ireland in 1815 up to her marriage in 1824. In his *Diary*, Richard Boyse, her disgruntled relation who complained that Mrs Hall's step-grandfather, George Carr, had squandered his, Richard's, patrimony, states firmly that Carr paid for Anna Maria's schooling in England.[35] Certainly, the account of school life in *Marian* rings true and we already know that Mrs Hall was at her best when speaking of her childhood days. She would have been old enough to have spent a year or two in a boarding school, and it is possible that the experience was so unpleasant that she never alluded to it openly in the nostalgic sketches of her youth, but detached herself and treated it purely as fiction. *Marian* is the only one of her four English novels, written between 1830 and 1840 wherein Mrs Hall shows her skills and in spite of the inherent absurdity of the plot, which relies on the old device of substituting one infant for another, the narrative proceeds smoothly, and with a certain amount of dramatic tension. There are some shrewd observations on folly and fashion in social life, and a fair sprinkling of humour, but the most interesting feature of the novel, apart from the satirisation of Lady Morgan as Lady Babs Hesketh, the literary lioness who played the harp for an enraptured social gathering, is Mrs Hall's attack on the private schools for girls which were then flourishing in England and the lesson which she wished parents to learn from it. Little Marian was sent as a boarder to a 'female seminary' when her adoptive mother grew tired of her, and although Miss Womble's establishment was not a Dotheboys Hall it was mean, cheerless and totally inadequate as an educational establishment, where learning was made 'a tax and a punishment'. In a footnote Mrs Hall referred to *Nicholas Nickleby*, published in 1838–39, saying: 'It may be necessary to observe that this work was written and in the hands of the publisher upwards of a year ago. Mr Dickens has since directed public attention to the "system" which too frequently ruins both mind and body, at ill-conducted schools'.³⁶ By so doing he had 'performed a duty for which humanity is his debtor'. Mrs Hall did not believe that 'atrocities' such as he described were common in the schools for "Young Ladies", a view which was not shared by Charlotte Brontë when she came to describe Lowood seven years later. There is no comparison between Mrs Hall's literary skills and those of either Charles Dickens or Charlotte Brontë but Mrs Hall's enthusiasm for a cause carries her along through a vivid description of the sad, lonely and restricted life led there by Marian, and she concludes:

> "God preserve our girls" ought to be the English parent's prayer until a totally different system of female education is established among us! The mania that possesses many rational persons in middle life to send their young daughters from the comfortable homes to a third or fourth-rate starving and perverted academy that they may imbibe a little bad French, and a little tuneless music, which is of no earthly use afterwards is truly a matter for marvel; still more wonderful is it that in these 'reforming days' the legislature does not enact some law by which persons should be examined to ascertain if they are in every way qualified for the task of instruction, before they are permitted to open schools.³⁷

Passages such as this can be written only from conviction and the strongest conviction is that which springs from first-hand experience.

Marian was very well received by critics in England, and even the Irish Nationalist periodical, *the Citizen* had high praise for the novel, commenting

> Mrs Hall has of late days occupied herself in endeavours to reform (?) the Irish people and in doing all that in her lay to make them English [a clear reference to *Stories of the Irish Peasantry*]. Perhaps, finding us too intractable, she has, at least for a time, abandoned her project, and leaving political economy to Miss Martineau, has turned to the more womanly study of the economy of the heart. The result is one of the best fictions of the season.³⁸

Admirers of Mrs Hall's work were promised a new publication from her pen in the year 1841. The portrait of Mrs Hall in the *Dublin University Magazine* concluded with the announcement that Mrs Hall, in partnership with her husband, was about to publish a work 'on the scenery and character of Ireland'. The Halls were 'intelligent authors' and could 'fairly anticipate a most liberal share of public favour for their undertaking. The work would be 'neither a guide-book, a tale, a history, or a book of travels, but [would] ... contain instruction for the tourist, amusement for the novel reader, information for the student, and novelties for the curious'. *Chambers's Journal*, ignoring Mr Hall's proposed contribution, pointed out that as 'Ireland was undergoing a rapid change for the better in its social aspect – it could not have a more zealous and impartial visitant than this excellent lady'.[39] Mr Hall was not to be eclipsed, however, by his wife's fame, and Halls' *Ireland; its Scenery, Character, etc.* placed him firmly in the limelight for the first time.

SEVEN

Halls' Ireland – 'Guidance for those who design to visit Ireland

PART I

'We have no design to write a guide-book;' said the Halls in their introduction to the chapter on County Wicklow in *Halls' Ireland*, 'although our leading object will be to offer some observations for the guidance of those who design to visit Ireland.'[1] The 'some observations' were published by How and Parsons as a part-work during 1841, 1842 and 1843. The complete work was published in a three-volume edition in 1843 and was most handsomely produced in a dark-green binding embossed with golden shamrocks and harps. Leading artists of the day, including Brooke, Maclise, MacManus, Nicoll, Prout and Weigall, contributed illustrations, and the engravings, etchings and wood-cuts were all of high quality. Each volume ran to over four hundred pages and every county in Ireland was described at greater or lesser length by Mr and Mrs Hall. The couple had together made five tours of Ireland, starting in 1825, and could claim first-hand knowledge of the country. To this they added the fruits of their researches into every aspect of Irish life – history, politics, economics, and culture in all its forms. Almost every chapter opens with a statistical survey of the county, and figures are given of acreage, both cultivated and uncultivated, and of population density. These were abstracted from the census of 1831 – a straightforward enough undertaking but one which was time-consuming. Then for the history of each county and its associations the Halls consulted every authority they could find. Some were well-known, some obscure.

Fynes Moryson, Giraldus Cambrensis, Holinshed, Geoffrey Keating, Cormac Mac Cuilleanain and Thomas Moore were joined by local historians, Dr Smith, the Reverend Mr Ryland, Mr Fitzgerald and Mr McGregor, Mr Windele, Mr Derrick, and Dr Woodroffe, among others. The Halls read Mr McSkimmin's 'History of Carrickfergus', Reverend Robert Walsh's story of New Geneva in Waterford, Father O'Daly's *History of the Geraldine Family*, Sir

Richard Musgrave's account of the 1798 Rebellion in Wexford (dismissed as 'an untruth'), Dr Stephenson's *Brief History of Grey Abbey in County Down*, and 'Memoirs of the Grace Family', a privately-printed paper by Sheffield Grace, F.S.A. In addition to the census returns they searched through statistical surveys of Antrim, Kildare, Meath, Cork and Roscommon for relevant facts and figures. For information about fisheries they read the reports of the 'Committee of Inquiry into Deep Sea Fisheries', for mining they consulted reports from the Mining Company of Ireland, the Knockmahon Copper Mining Company, and the leadmines of Lurganure. The 'Report of the Commissioners on Municipal Corporations in Ireland', the 'Report of the Select Parliamentary Committee of 1823 on the Condition of the Poor in Ireland', and reports from the Poor Law Commissioners gave information about social conditions in the country and these were supplemented by a report from the Loan Fund System and by returns from Savings Banks. Then there were accounts of tours by earlier travellers, books and papers by naturalists, and the many antiquarian records by such authorities as Croker, Grose, Petrie and Betham that furnished more details of the rich variety of Irish life, past and present.

There is no doubt that the Halls prepared their work very thoroughly, and that, as they said themselves, they 'laboured with zeal and industry to obtain such topographical and statistical information as might be useful to those who visit Ireland'.[2] However, Mr and Mrs Hall were successful and experienced journalists, and knew what would and would not appeal to the public. A purely factual account of the state of Ireland, historically, politically and socially, would not be popular and would not sell enough copies to make publication worthwhile. With all their protestations, genuine though they probably were, of wishing to 'make Ireland more advantageously known to England', the Halls had a sound commercial reason for writing a popular book about Ireland – it would make money. Mrs Hall had already discovered this and had become established as a writer who was an expert on Irish life and Irish customs. Her skill in depicting native characters and her talent for telling old Irish tales could, therefore, be put to good use in the new work without in any way compromising its seriousness. According to an advertisement for the series in the *Literary Gazette* in 1840 the 'more important details' (the facts and figures) would be joined 'such incidents, descriptions, legends, traditions, and personal sketches, as might serve to excite interest in those who are deterred from the perusal of mere facts, if communicated in a less popular form'.[3]

That is an accurate description of Halls' *Ireland, its Scenery, Character, etc.* – a mixture of statistics and stories, of fact and fiction, of history and legend. It is a very uneasy mixture, as one might expect, and the transition from one mode to another is almost always an awkward and abrupt one. Yet the work is intensely interesting, not least for the light it throws on the Halls themselves, especially Mrs Hall. She had great qualities of endurance and physical stamina, as well as an emotional resilience that sustained her in difficult circumstances. When the young newly-married Halls first travelled to Ireland in 1825 it was just before the introduction of the steam packets between Ireland and England. 'The voyage across was', they recalled in the opening paragraph of *Halls' Ireland*, 'a kind of purgatory for the time being, to be endured only in cases of absolute necessity.' The discomfort of a cramped cabin on board a schooner, the lack of food, the poor quality of service, the miseries of sea-sickness, these were minor matters compared to the uncertainty that was the lot of those who were at the mercy of the wind and the weather. A voyage should take only three or four days, but, recalled the authors, 'it was once our lot to pass a month between the ports of Bristol and Cork; putting back, every now and then, to the wretched village of Pill, and not daring to leave it even for an hour, lest the wind should change and the packet weigh anchor'.[4]

The coming of steam made the sea voyage an 'excursion of pleasure', but travel on land still had its share of discomforts and difficulties. In areas not served by the Bianconi coaches (introduced in 1815 and travelling through nearly every district of the south of Ireland. 'Persons of the highest respectability' used them regularly, for they were comfortable as well as safe), it was necessary to hire cars and only too often they were 'badly built, dirty and uncomfortable'. They were not always available when required, and if the one car of the neighbourhood was out, then the traveller had to wait for its return. The Halls advised the tourist 'to lay in a stock of good humour – for petty annoyance will frequently occur' and added that 'a plentiful supply of waterproof clothing was also essential 'for sunny June is no more to be trusted than showery April'.[5] This advice was repeated in the chapter on Galway. The traveller should provide 'amply against the weather. A Connemara shower is like the descent of an avalanche of water and drenches everything in less than a minute. Umbrellas are perfectly useless; the hill blasts tear them to shreds before they can be raised. The wind rushes so fiercely down some of the passes that our horses found it impossible to progress faster than a mile an hour'.[6] The unreliability of the Irish weather was 'a sad drawback upon pleasure' but it was an evil that could be guarded against, and the Halls were quick to find ways of enjoying their trips. The covered car, for example, was not only stuffy and confining but prevented travellers from seeing the beauties of the countryside. There was nothing the Halls could do about the design of the car, but they remedied 'the evil of confinement' by stopping at every promising spot, and either getting out or making the driver turn his vehicle around so that they could view the scenery from the rear window, with its oilskin curtains tied back. As far as they were concerned the wit and humour of Irish cab drivers made up for any deficiency in their vehicles and compensated for some of their own shortcomings. These 'thoughtless and reckless' men who 'lived upon chance, always '"taking the world aisy"', were the exact opposite of English postillions. The Englishmen did their duty and were respectful but they were impersonal, and there was no warmth or friendliness in their dealings with the traveller. Irish drivers, on the other hand, were inquisitive, always wanting to know your business, and ready to join in your conversation, be it general or not. They could never be accused of actual impertinence, but they came very close to it. Nevertheless, the Halls were happy to chat with them and valued their friendliness. They were not, however, enthusiastic about another mode of travel – that by canal. It was by no means pleasant for the boat was 'exceedingly narrow' and the passengers were 'painfully cramped and confined'.

Most of the inns the Halls lodged in were reasonably clean and comfortable but occasionally they chanced on a bad one. There was one in the west of Ireland in a 'miserable village' where the Halls spent a wretched night or, rather, as they recalled

part of a night, for we rose from our beds an hour before daybreak and pursued our journey. There was neither tea nor bread to be procured: the horse, the cow, the pigs and the hens were separated from us by a floor, through the divided boards of which they had the ample opportunity for conversing with us – which they did not fail to do. Soon after midnight, our domicile was invaded by the hostess, who required from the cupboard some "refreshment" for his reverence, who had just arrived from a station, and about an hour afterwards the corn-bin was to be applied to for "a feed" for his reverence's pony, who had to make a new start. This break-in was followed by another; the "boy" wanted his "top-coat", for the rain was "powering down;" a short while afterwards the household was all in motion, and our chamber contained everything that was wanted.[7]

The tone is one of wry humour, not of severe criticism. In another inn in County Cork the Halls were bitten by fleas and when they showed the angry red marks to the serving girl they were completely disarmed by her quick-wittedness. She looked at the spots and exclaimed 'in a tone of mingled anger and repugnance, "Why then, bad luck to the dirty bastes at the house ye *last* slept in"'.[8]

The only strongly-worded complaint the Halls made about the lower classes of Irish people they met on their travels concerned the guides at Glendalough. They simply would not leave the tourists alone to absorb the beauty of the scene and to savour its atmosphere, but followed them relentlessly. 'You give up all thought of quiet, in despair. Guides of all degrees start from beneath the bushes, and from amid the crags – we had almost written, from out of the lake – and "they will do anything in the wide world to serve and obleege yer honours," except leave you to yourself.' They were persistent and intrusive, but the Halls noted their honesty, their civility and their peaceful behaviour. In spite of these undoubted virtues the Halls wished them ill in an untypical outburst: 'For ourselves, we confess a strong desire to sink the whole tribe, male and female, into the deepest pit of the deep lake'.[9] This reaction was in sharp contrast to that evoked by the guides at Killarney. They also were very numerous and very vociferous, but when once a candidate had been chosen from the throng he was an excellent companion. No praise was too high for these guides:

They are the most amusing fellows in the world; always ready to do anything, explain any matter, go anywhere – for if the tourist proposes a trip to the moon, the guide will undertake to lead the way – "Bedad he will, wid all the pleasures in life". They are invariably heart-anxious to please; sparing no personal exertion; enduring willingly the extreme of fatigue; carrying as much luggage as a pack-horse; familiar, but not intrusive; never out of temper; never wearied of either walking or talking; and, generally full of humour. They enliven the dreariest road by their wit, and are, of course, rich in old stories; some they hear, others they coin, and, occasionally, make a strange hodge-podge of history – working a volume of wonders out of a solitary fact. If they sometimes exact more than is in "the bond," they do it with irresistible suavity.[10]

All the guides found the Halls to be adventurous tourists. If not exactly reckless, they were almost indifferent to possible danger. At the Old Weir Bridge in Killarney, between the Upper Lake and the Torc Lake, where the current is quick and strong, travellers usually left their boat and walked across to embark again on the other side, for the passage was considered rather dangerous. Mr and Mrs Hall, however, stayed in the boat and 'shot through with frightful rapidity'. It was evident, they added, 'that a very small deviation either to the left or to the right would have flung us among the breakers, the result of which must inevitably have been fatal'.[11] The onlookers, who were 'watching with some anxiety' cheered the pair when they were safely through, and congratulated them on their courage. At Glendalough the Halls climbed up to Saint Kevin's Bed and noted that the ascent was 'exceedingly difficult, and somewhat dangerous, for a slip would inevitably precipitate the adventurer into the lake below'.[12] The couple regularly travelled along narrow paths 'at the brink of precipices', and took several boat trips among mountainous seas in order to see for themselves what earlier travellers had only been told about by others. At Killarney the Halls persuaded workmen to remove large stones in order to reveal the entrance to a cave. The opening was 'wide enough to admit the body of a man' and Mr and Mrs Hall entered in. They went down a slope, when some 'loosened stones' fell, and informed them that there was water below. Having tested its depth with their walking sticks they realized that it was only about two feet in depth so they went on and found themselves 'in a cave ... about sixty feet in diameter and five feet in height'. 'Peering narrowly about' Mr and Mrs Hall spied 'a hole that looked like a fox-hole. It was, however, barely big enough ... to crawl through' but they persevered and found another cave, and then another.

In Tipperary Mr and Mrs Hall visited the Mitchelstown Caves, which had been discovered in 1833 and were already famous – ('a natural marvel, the most singular in Great Britain') – but not very easily accessible. When the Halls entered they first went through a 'narrow, gradually sloping passage of about four feet in height' terminating in 'an almost vertical precipice, about fifteen feet deep', which was descended by ladder. Elsewhere in the caves they descended the 'chimney', a work of some danger, noted the Halls, for it was 'barely wide enough to allow a passage', and its sides had 'very few projections on which to place the feet'. A guide, however, went 'before the visitor, directing his steps and frequently giving the foot a resting place upon his shoulder'. A slip would have been fatal, as the chimney descended to a depth of ninety feet. Finally, the party was ushered by their guides through a burrow which was so narrow that they had to twist themselves along it, 'after the fashion in which the screw makes its way into a cork'. The Halls recalled their progress:

The task required physical strength, and no inconsiderable nerve; for the passage extended at least one hundred yards, the greater portion of which was necessarily traversed by crawling through a space barely two feet square, sometimes so reduced as to render indispensable the kind of "twist" we have referred to, and repeatedly suggesting the painful suggestion that a fall of two or three inches, in any of the rocks above or around us, would enclose us prisoners beyond the possibility of rescue.[13]

The entire venture took five hours, all of them spent underground, and they covered about eight miles in all. The Halls greatly enjoyed the trip and when they visited the tumulus of New Grange in County Meath they made a similar expedition, not as extensive, but equally difficult. 'We crept, or rather crawled', they wrote

along a distance of about sixty feet, the height being no more than eighteen inches and the breadth somewhat less than twenty-four feet. The passage is roofed and the sides are supported by enormous slabs. About midway there is a stone which appears to have fallen from the perpendicular and seems to forbid further progress, but this was passed, however, by twisting the body onwards.[14]

All this was done in total darkness, before a supply of candles was brought in to illuminate the central chamber.

Although Mr and Mrs Hall did not wish their work to be thought of as mere 'guide-books', the books have some practical advice for the tourist. It is not only the major and obvious sights that are described and recommended to the traveller's attention, but lesser-known and obscure ones. The couple regularly turned aside from the main roads to point out places of natural beauty, 'small unrecorded places – nooks hid beneath cliff or mountain', or interesting and historic ruins that might otherwise have been overlooked. As they pointed out, 'a mile or two of wandering off the beaten track' will often bring rich rewards to the traveller. 'Let him', they said, 'make up his mind to loiter.' From Clifden to Leenane in County Mayo, for instance, the coach road runs for a distance of twenty miles, but the Halls urge the tourist to make it a thirty-mile journey by verging to the right to visit 'the graceful shores of Renville and the rugged pass of Salruc'. To do this, and to experience the wonder of the magnificent scenery, it will be necessary to hire 'a stout horse and a strong car', but the trouble will be well worth it. Similar advice is given to the traveller in County Antrim who is on his way to see the Giant's Causeway. If he turns aside from the high road from Belfast and visits the peninsula of Island Magee he will find that 'its scenery, bays, highlands and caves are highly interesting, independently of the objects it presents in great variety to the geologist, naturalist, and, in some degree, to the antiquary'.[15]

Mr Hall, as an antiquary (elected a Fellow of the Society of Antiquaries in 1842, proposed by T. Crofton Croker and seconded by Sir William Betham), was anxious that the pre-historic site at Lough Gur in County Limerick should be seen by all who travelled in that region. It 'claims particular notice at our hands', said he, 'because it has received so little attention from previous visitors and even from the county historians'. He considered that the 'extensive assemblage of druidical remains' around the lake and on its principal island made it 'the most interesting spot in Ireland for the visitor in search of the unique, and yet, strange to say, these gigantic relics, which extend over so many miles of country, have been allowed to remain unexplored and undescribed'.[16] Conversely, he was often pained by the neglected aspect of so many historic ruins – or worse, by the evident depredations of the peasantry who often used ruined churches as burial places and made headstones of 'fragments of broken pillars, mullions and fretted work'. In Kilkenny, a city which was full of 'striking, interesting, and ... picturesque ruins' it was distressing to find 'wretched hovels propped up by carven pillars' and in several instances the Halls 'discovered Gothic doorways converted into entrances to pig-styes'. Both Mr and Mrs Hall, however, found great pleasure and interest in viewing domestic artefacts which had not changed in design over the centuries and were in daily use in peasant cottages – comfortable ones, that is. In a little house in County Galway where Mr and Mrs Hall had taken shelter from the rain they noted a 'singularly primitive chair, very commonly used throughout Connaught ... roughly made of elm, the pieces nailed together'. This piece of furniture, they stated, had undergone little change during the previous eight or ten centuries. In the same cottage they were lucky enough to see what was rapidly becoming a rarity, a quern or handmill. This too was of ancient design, as was the wooden drinking cup – 'a modern substitute for the mether'.

In the course of their travels Mr and Mrs Hall were able to stay with some of the many friends they had throughout the country, but they 'studiously avoided all reference to the seats or domains of country gentlemen – except where improvements carried on in particular places excited and deserved comment'. Their particular friends, the Grogan Morgans of Johnstown Castle were admiringly mentioned in the chapter on County Wexford, but that was in reference to their good work as resident landlords. 'We have abstained', said the Halls, 'from intruding our own personal acquaintances upon the notice of the reader ... for we should have ill requited kind and gratifying attentions, if we have made private individuals topics of public conversation.' They made an exception though, for Miss Edgeworth, for, as they pointed out, 'Edgworthstown ... may almost be regarded as public property. From this mansion has issued so much practical good to Ireland, and not alone to Ireland, but the civilised world – it has been so long the residence of high intellect, industry, well-directed genius and virtue, that we violate no duty by requesting our readers to accompany us hither'. They then described their visit in some detail; the layout of the house, its furnishings, the daily routine, and, of course, the sayings and the doings of the famous Miss Edgeworth – that 'unspoiled and admirable woman'. This portion of the work is undoubtedly written by Mrs Hall, not only is the style hers – that blend of attractive simplicity and rather sententious moralising – but it refers to the part Miss Edgeworth played in Mrs Hall's own career as a writer. 'If we have ourselves', runs the account, 'been useful in communicating knowledge to young or old – if we have succeeded in our hopes of promoting virtue and goodness – and, more especially, if we have, even in a small degree, attained our great purpose of advancing the welfare of our country – we owe, at least, much of the desire to do all this, to the feelings derived in early life from intimacy with the writings of Miss Edgeworth.' These are Mrs Hall's sentiments, ones which have already been voiced in her works. The paragraph concludes with conclusive proof that this portion of the work is from the pen of Mrs Hall for it adds, 'Much, too, have we owed to this estimable lady in after life. When we entered upon the uncertain, anxious, and laborious career of authorship, she was among the first to cheer us on our way; to bid us "God speed"'.[17] We know that one of Mrs Hall's early books (The Second Series of *Sketches of Irish Character*) was dedicated to Miss Edgeworth, and Mr Hall in his autobiography, *Retrospect of a Long Life* refers to the encouragement Mrs Hall received from the older, more established author:

When Mrs Hall published her *Sketches of Irish Character* she ventured ... to send a copy to her renowned countrywoman: she received in reply a thorough analysis of the book, a note upon each and all of the stories with very warm praise of the whole. There was not only no tone of jealousy, there was a strongly expressed joy that another author was rising to continue in a safe, right and holy spirit the work Maria Edgeworth had done for Ireland.[18]

He quotes too, a letter from Mrs Hall in reply to one of Maria Edgeworth's that 'argued strongly for truth in fiction'. Mrs Hall wrote, 'I ventured, not withstanding my homage for that estimable woman, to ask her if her portrait of Sir Condy, in *Castle Rackrent* was a veritable likeness, and endeavoured to convince her that to call imagination to the aid of reason – to mingle the ideal with the real – was not only permissible but laudable as a means of impressing truth'.[19] According to Marilyn Butler, Miss Edgeworth's biographer, the older and more famous writer greatly enjoyed Mrs Hall's *Sketches of Irish Character*, referring to it in a letter to one of her friends as 'touching – simple, beautiful and true Irish'.[20]

Mr and Mrs Hall were aware that not all prospective tourists in Ireland would be able to stay with hospitable friends and they refer quite often to inns and hotels which can be recommended to travellers. In Cork city the Imperial Hotel could 'vie for elegance and comfort with any hotel in the kingdom', the little inn at Dunmanway was 'very clean and, considering all things, comfortable', and that at Macroom looked 'set to prosper', while the Royal Victoria Hotel at Killarney, kept by Mr Finn, was 'a very splendid establishment' that bore comparison, both inside and out 'with any hotel in Brighton or Cheltenham'. There was extremely high praise for White's Hotel in Wexford – 'one of the best, if not the very best hotel in the south of Ireland'. 'We have never visited a better-managed establishment' said the Halls. They added, however, with scrupulous honesty, that they had never *stayed* there themselves, having had good friends in the neighbourhood, but that they had 'received the highest testimonials from many as to the cleanliness, order and attention of the house – and especially in reference to Mr White's cuisine'. In Roundwood, County Wicklow there were two good inns, Heatleys and Murphys, both of which were 'exceedingly clean and comfortable', as was the Wooden Bridge Inn, where the 'charges for entertainment' were exceedingly moderate. Booking ahead was advisable here, as the hotel became very crowded in the season. Near Rathnew there were two inns of note, one at Ashford, and one at Newarth Bridge. The Halls strongly recommended the latter for 'all matters about their hotel are neat, clean and well-ordered and nothing can exceed their attention to their guests'. Charges there were very moderate.

Dublin was well supplied with hotels 'in all the fashionable streets' and the most popular perhaps was Greshams in Sackville Street, but the Halls remarked that 'the old establishment of Morisons' retained its reputation for comfort, attention and moderate charges. In the north of the country the Halls were not so lucky in the matter of accommodation. They were very happy with their hotel in Belfast – the Donegal Arms – which was a 'well-ordered ménage', with a 'courteous and inquiring landlord, exceedingly attentive servants and good posting', but they could not say as much for the hotels of the north in general. An exception was Miss Henry's place at Bushmills, 'neat, well-ordered and comfortable', which the Halls recommended 'in the strongest terms'. The west of Ireland was well provided with inns, and that of a Mr Anthony O'Reilly at Leenane, County Galway, so impressed Mr and Mrs Hall that they reprinted a copy of his prospectus in their book. Usually no details of prices were given, the Halls merely saying that charges were moderate, but the value for money they received in the Mucross Hotel at Killarney so impressed them that they itemised the 'absolutely startling' charges:

Two breakfasts	3s.0d
Two dinners	4s.0d
Pint of wine	2s.0d
Two teas	2s.0d
Bed	1s.6d

'The moderate tourist', they added, 'will calculate his necessary expenses at Killarney at something less than seven shillings per day. The only charge at which he will complain is the hire of a boat – sixteen shillings – but it includes the dinners of five men. The charge for a pony to Mangerton or "The Gap" is five shillings.'[21]

Only one complaint was made by the Halls about charges in Ireland – they considered that 'the cost of travel by outside jaunting car in County Wicklow was "unreasonably high". They were most emphatic, however, that a trip to Wicklow should not be missed by intending travellers. 'Nowhere, perhaps, in the world' said the Halls, 'can they be so largely repaid for so small an expenditure of time and money. A journey of twenty-four hours [from London] may place them in the centre of it. A railway carriage conveys them to Liverpool; the steam-boats – the largest, safest and best in the kingdom, which ply twice a day – in little more than ten hours to Dublin; and Dublin is within an hour's drive of the country.'[22] In Wicklow the tourist could gaze on the sublimity of the mountains, their wild bleak grandeur, and the 'noblest of magnificent cataracts'. Descriptions of scenery take up a large part of *Halls' Ireland*, and certain phrases recur in the text. Favourite landscapes are of 'exceeding wildness and singular beauty', flowers are 'a mass of living gold', glens are 'deep and richly-wooded', hills are 'precipitous', rocks are 'black and savage', islands are 'gracefully wooded' or are 'luxuriantly clothed in the richest verdure and foliage'. It is savage beauty that most appeals to the Halls, however, and Gougane Barra in County Cork, the Giant's Causeway in County Antrim, The Gap of Dunlow in County Kerry and the mountains of Connamara evoke long and ecstatic outpourings of praise. In common with their contemporaries the Halls admired both the sublime and the picturesque, and found many a moral lesson in nature. Ireland, they pointed out, was a painter's paradise, Connamara in particular, was rich in subjects for an artist's pencil:

> The artist by whom this district has not been visited, can indeed have no idea of its surpassing grandeur and sublimity – go where he will he finds a picture; the lines of the mountains covered with heather; the rocks of innumerable shapes; the "passes" rugged, but grand to a degree; the finest rivers, always rapid – salmon-leaps upon almost every one of them; the broadest and richest lakes, full of small islands, and at times dotted with luxuriant foliage along their sides; in fact, nature nowhere presents such abundant and extraordinary store of wealth to the painter – and even now it has been very little resorted to.[23]

Why, ask the Halls, do English painters go to France and Italy or to the 'hackneyed' and 'sodden' Rhine in order to find suitable subjects for their art? Ireland is much nearer, equally blessed with magnificent scenery. They quote some remarks by the artist Mr Fairholt (who accompanied them on an Irish tour) which were published in Mr Hall's journal the *Art Union* in support of their contention that Ireland is more rewarding to the painter than is the Rhineland. Mr Fairholt was very forthright. 'The Rhine', said he, 'has been exhausted; its scenery has been copied and recopied until it has become so familiarized as to be almost looked on with indifference.' Artists had often ventured far afield, with risks to their health, even to their life, in search of new ground, while near to home – 'indeed part of Great Britain' – there was an unspoiled almost unknown country of surpassing beauty.[24] It was not only the 'glorious mountains' and the 'lovely glens' that would delight the artist; the inhabitants of this region of natural beauty were themselves fit subjects for the painter. The girls 'in their deep red petticoats and jackets' had an antique grace and dignity, and were as picturesque as the cottages in which they lived. As the Halls pointed out, the peasants were usually seen in groups, and formed a 'valuable accessary [sic] to the landscape'. This comment is a direct echo of William Gilpin's advice to artists who were sketching landscape; 'In *adorning your sketch* a figure or two may be introduced with propriety. By figures I mean moving objects, as waggons and boats as well as cattle and men ... Their chief use is, to make a road – to break a piece of foreground – to point out the horizon in a sea-view'.[25]

Figures were still subordinate to landscape in the artistic sense, and the Halls' book, in spite of its detailed descriptions of certain individuals' appearance and behaviour, and its account of society in general, does leave an impression of a country rich in natural beauty, but suffering from the backwardness of its population. This impression is the result of the Halls' literary limitations, and their innate prejudices, for it was certainly not due to a lack of diligence. The couple examined every aspect of Irish social and economic life, having already studied what authorities they could find. They personally inspected institutions such as jails, workhouses, schools and penitentiaries, and visited collieries, factories, hospitals and the barracks of the constabulary. Botanic gardens, colleges of agricultural education, the new Roman Catholic College at Maynooth, historical and philosophical societies in Dublin and in Belfast, all were visited, studied and written about. Wherever possible, the Halls went to see things for themselves. Their researches were usually scholarly and infused with a high moral purpose, but one of their investigations was delightfully out of character. Having often heard about the extensive smuggling that was carried on from the coves and inlets of the southern and south-western coast of Ireland, Mr and Mrs Hall persuaded a friend of theirs who was engaged in the trade to allow them to accompany him to receive the cargo as it came ashore. This was the most hazardous part of the enterprise. Mr and Mrs Hall cautiously followed their guide 'through dells and precipices' to the shore which was 'literally covered with men and horses'. The couple were not content to wait tamely there but climbed aboard one of the boats that was rowed, with muffled oars, out to where the lugger and its contraband cargo was waiting. A courteous captain offered refreshments in his cabin but the pleasant party was soon interrupted by the news that a vessel was approaching the lugger. The alarm was given, and the crew – 'a motley group composed of the hardy and desperate of all nations – stood by, arms at the ready, prepared to repel the expected attack'. It was an awkward moment for the Halls who were, they recalled, 'apprehensive that we might have to pay a frightful penalty for our curiosity; for when reflection came, it came too late; we had no means of returning to land, and were compelled to share the destiny of our comrades of the moment, whatever that destiny might be; the easiest, perhaps, a trip to Holland'.[26] Fortunately for the Halls, (and for the smugglers), it was a false alarm, and the venturesome pair returned home safely. This incident had taken place in the 1820s, and presumably the passing of time had given it a gloss of respectability, but the relish with which the tale was told by the Halls is at odds with their otherwise bland respectability. The narrative is brisk, the prose lively, and the tension is high. Then, as if repenting of the frivolity the Halls write soberly in the following paragraph of the text that 'the curious in such matters may now examine, all along the coast, numerous holes and caves, formerly depositories of smuggled goods', and may visit one near Glengariff, known as Brandy Island 'stories in connexion with which will be related to him, in abundance, by the boatmen'.

The Halls seem to have been interested in almost everything and to have assumed that their readers would share their enthusiasms. Only an optimist could have imagined that the intending tourist would wish to read over twenty pages on the linen industry with detailed accounts of the methods used to convert the blue-flowered flax plant into a damask tablecloth. This is the most extreme example of the Halls' essays on production, but there is also a wealth of information on how to pan gold, cut turf, dig coal and fish for lobsters. In some parts of the country if the authors felt that the natural features and the history of the area were lacking in interest they filled up the space with an essay. County Cavan, for instance, possessed 'no feature of a striking or peculiar character', so the Halls did not 'detain the reader in this comparatively uninteresting county', but 'took the opportunity of introducing some general remarks' on the subject of the Irish language. County Waterford is dismissed as 'the least interesting, and certainly the least picturesque, of the counties of Ireland', so there is an excuse to present a long essay on holy wells and the 'revolting' superstitions connected with them, and another on the characteristics of Irish waiters. County Carlow 'possessed no feature of a peculiar or exclusive character', but the Halls were happy to educate their readers in the workings of the flour mills in the county. County Westmeath, too, was lacking in any remarkable features, so the Halls presented their readers with 'some information concerning Irish music'. In the chapter on County Monaghan they related 'some anecdotes illustrative of the "good people" [the fairies]' because the district yielded 'but a scanty supply of materials of an original character'.

It is not only in these shorter chapters that the Halls introduced informative essays – they are to be found throughout the three volumes of the work. These essays are on topics that the Halls considered to be of 'grave importance' including round towers, the Irish language, education, faction fights, landlords, secret societies, Irish music and religion. The personal views put forward are, in many cases, those already expressed by Mrs Hall in her Irish stories and sketches. On religion, for instance, the tolerance she earlier displayed is once more evidenced in her assessment of the part that could be played in Irish society by Roman Catholic priests who were educated to be 'liberal, enlightened and charitable men'. They would do much to dissipate the fear and distrust of priests that was felt by those 'unwise, unchristian and intolerant sectarians' who could see no good in Popery. At the same time, she voices yet again her suspicions of the new breed of Maynooth-educated priests who were utterly unlike the old breed of priests she had known in her young days in Bannow. Such a priest, having been educated abroad, was 'an accomplished gentleman, benevolent, courteous and charitable'. 'The Maynooth priest, she stated, was 'of another stamp ... of humble birth and connexions' and utterly ignorant of society. She feared that the training at Maynooth would not 'remove all that was objectionable in his previous habits and education'[27] but would merely strengthen it. Nonetheless, she was not one of those who wished to see the government grant to Maynooth withdrawn – on the contrary, she would like to see it augmented, so that a decent wage could be paid to the professors of the college (whose salaries were 'scarcely enough to pay a stonemason') and that standards could be improved.

From the very start of her writing career Mrs Hall had emphasised the duty that landlords owed to their tenants and the desirability of their residence on their Irish estates. In the three volumes of Hall's *Ireland* the theme is addressed again and again. Good landlords are often cited by name, including the La Touches, and the Putlands in Wicklow, the Grogan Morgans in Wexford, the Kavanaghs in Carlow and Lord Glengall in Tipperary. These are all resident, but the Halls admit that even where a landlord is an absentee a good agent can look after the interests of the tenants. Captain Pitt Kennedy, agent for Sir Charles Style, the owner of extensive estates in County Donegal, was held up as an example to others in the same managerial position, for by his own imaginative efforts and the enthusiastic support of his employer he transformed the property and the lives of the tenants who would otherwise have had to choose between emigration and starvation:

'Now', record the Halls, 'they are prosperous, industrious and happy. Where the foxes of the earth made holes, their cottages are built; land, over which the screaming eagle flapped its wings, echoes with the hum of cheerful voices. Children, ignorant of all book-knowledge, and wandering like Indians over the hills and valleys, are gathered in the profitable union of a happy school, and taught the independence produced by steady and well-directed labour.'[28]

Land has been reclaimed, and tenants are cultivating their own plots while at the same time improving the value of their landlord's property, and an agricultural school is educating the younger generation of farmers in new methods of cultivation. This is what can be done when a proprietor has the best interests of his tenants at heart and does not leave them to the mercy of an unscrupulous agent or middleman. Mrs Hall had often written scathingly of middlemen – a race she truly despised – and while the Halls believed that as a species they were disappearing from the country, they took the opportunity to tell English readers about their origin and behaviour. They were the 'evil productions of an evil system', charged by an absentee landlord with the duty of extracting rents from tenants, and enriching themselves in the process. The authors explained how this was done. The middleman parcelled out the estate in small bits and let them out annually, never on lease. These slips of ground seldom produced anything other than potatoes for the family, and for the invaluable pig. The land was let at the highest possible rate, and to the highest possible bidder. So great was peasant land hunger that in many cases tenants contracted to pay ludicrously uneconomic rents, were soon in arrears and eventually failed to meet their legal requirements and were evicted. At all times they were in the power of the middleman, who had golden opportunities to enrich himself. The Halls described the system and its consequences simply and clearly, and a thoughtful tourist who wished for an explanation of the poverty and misery which he saw in many parts of Ireland would find it in their words.

In *Sketches of Irish Character*, in *Lights and Shadows of Irish life* and in *Stories of the Irish Peasantry* Mrs Hall had identified intemperance as one of the great flaws in the native Irish character and in the two later works had mentioned the early crusade against strong drink. Now, in *Hall's Ireland* she and her husband were happy to report on the work of Father Theobald Mathew, the Capuchin friar from Cork whose temperance movement was proving to be such a huge success. The effects of his campaign were immediately obvious to the Halls, and the subject 'was one of such vital importance' that they discussed it at length. Drunkenness had been the 'shame and bane' of Ireland and an Irishman had 'become proverbial for intoxication'. As far as the Halls were concerned 'all the mischievous tendencies of the lower Irish' might be traced to 'their habitual intoxication', and there seemed to be no hope that an Irishman could be weaned away from what was his only solace. There had been attempts to do so notably by Mrs Hall's own 'connection', the Reverend George Carr, but Father Mathew's movement was the first to have proved successful. The Halls admitted that they had at first been sceptical about reports of the friar's success, but they then saw for themselves the charge that had come about in Irish society. During a month's trip in the summer of 1840, spent in touring Cork and Kerry, on 'by-ways as well as highways, visiting small villages and populous towns; driving through fairs; attending wakes and funerals' they did not meet 'a single individual who appeared to have tasted spirits'. In the following three months, travelling in other parts of Ireland they saw 'but six persons intoxicated'. Of the fifty or so car-drivers they employed not one would accept a drink, and even the Killarney boatmen, 'proverbial for drunkenness, insubordination and recklessness of life' declined whiskey and refreshed themselves with lake water. From their own observations the Halls concluded that sobriety was 'almost universal' throughout Ireland, and adduced diminished Excise Returns as proof of their statement. The crime rate had fallen, and deposits in small savings banks had increased, both circumstances that could be traced to the crusade against intemperance. Mr and Mrs Hall had a strong personal admiration for Father Mathew, and spoke movingly of their meeting with him. He has been, they said, 'stimulated by pure benevolence to the work he has undertaken. The expression of his countenance is peculiarly mild and gracious: his manner is persuasive, gentle, simple and easy, and humble without a shadow of affectation, and his voice is low and musical – "such as moves men"'.[29] The Halls trusted that their testimony to Father Mathew's goodness, and the success of his campaign would be accepted by all, as they could certainly claim to be without prejudice, adding, 'our opinions, both religious and political, are certainly not of a nature to bias us unduly'.

The Halls did strive hard to be free of all bias and prejudice, whether religious or political, and their version of Irish historical events is generally even-handed. They deprecated the introduction and imposition of the iniquitous Penal laws that reduced Roman Catholics to a state little better than slavery, and they expressed admiration for a people who clung to their own religion in spite of persecution. They duly expressed horror when speaking of Oliver Cromwell's massacres at Drogheda and at Wexford, and they lamented the fate of the great and noble Irish chieftains who were dispossessed of their lands and who either fled to foreign lands or who lingered as beggars in their own. However, the sentiments of true blue honest Protestants come through in their accounts of the Siege of Derry and the Battle of the Boyne. The Protestant citizens of Derry, the Halls told their readers, were 'threatened by an undisciplined mob of armed men, recently recruited from classes where evil passions required no stimulus, and governed by rulers who made no concealment of a resolve to destroy their rights and their religion' and banded together to defend their city. This they did gallantly and their success led directly to the victory at the Boyne, the securing of the British Crown for the Prince of Orange, and 'sealed the charter of liberties obtained by "the Revolution"'.[30] Mr and Mrs Hall granted that it was only for Protestants that the memory was "glorious and immortal" but they trusted that in time, party spirit would disappear, and that the siege of Derry would take its place with the siege of Limerick in the history of Ireland as 'evidence of the courage, fortitude and endurance of which Irishmen are capable. Derry is the twin of Limerick; the sieges of both are alike honourable to the brave spirits who maintained both – the Catholic in the one case and the Protestant in the other'.[31] This authorial even-handedness is forgotten in the Halls' account of the Battle of the Boyne where the English King William, Prince of the House of Orange, led an army drawn from several European states, but Protestant as to religion, against the usurped King James the Second of England, head of an army that was composed of English, Irish and French troops, and staunchly Catholic in religious belief. This battle was, as the Halls point out, a decisive one, not only in Irish terms, but in European ones, because of the national interests involved. The victory for William's army ensured the success of Protestantism in England, and challenged the threatened French domination of Europe, and was, concluded the Halls, 'the key stone of the temple of civil and religious liberty in these kingdoms'. They preferred not to add that William's success there led directly to the imposition of the Penal Laws which deprived Catholics of civil and religious liberty but concluded their detailed account of the battle (complete with a sketch of the battlefield) and its happy consequences by musing 'It is no marvel, therefore, that the battle of the Boyne river is held sacred in the memories of all Protestants – those of Ireland most especially; and that, ever since, its anniversary should have been a season of thanksgiving and rejoicing'.[32]

PART II

The three volumes of Mr and Mrs Hall's huge work comprise more than a mere guide book for prospective tourists. In addition to the practical information useful to a traveller they present an historical and sociological account of the state of Ireland in the early eighteen-forties. A diligent reader would have been very well informed about conditions in Ireland before he even set foot in the country if he had taken the trouble to read through the whole work. However, as the Halls were well aware, hard facts and even strongly-voiced opinions were not enough of themselves to ensure reader interest – what was needed was something to catch the imagination. Stories, legends, personal reminiscences, these were what people liked to read, and Mrs Hall was the perfect writer to supply them. Her contribution to *Halls' Ireland* may be looked on as a continuation of her fictional writing career, and it bears all the marks of her style. She is concerned, as always, to give an insightful picture of Irish character. This time, however, in contrast to her earlier habit of emphasising both the virtues and vices of the Irish peasant's nature, she lays greater emphasis on the good points. She is not now, after all, seeking to redeem the Irish peasant; the work is not addressed to him, but to the English tourist who must be taught to appreciate the good qualities of his Irish neighbour. Hence the insistence on the improvements in Irish society that she has noted since she last wrote about it in *Lights and Shadows of Irish Life*.

Drunkenness has almost disappeared, and Irish peasants are now showing habits of economy. Faction fights, which were once a feature of Irish life, and which Mrs Hall and her husband had witnessed several times, were now rare occurrences. The couple visited Donnybrook Fair, once notorious for 'its dissipation and vice', and the ease with which it filled the jails with criminals, and found nothing there but good-humour and innocent merriment. There are still crimes of violence, but these are generally committed by members of secret societies such as the Whiteboys, the Ribbonmen, or the Shanavests and are directed against landlords. An ordinary tourist has nothing to fear in this regard, and may journey without fear of injury to his person or his property. Mrs Hall had always expressed a detestation of violence, even though her stories and sketches are often based on acts of violence, and most of the comments in *Halls' Ireland* on the activities of the secret societies are initially what might have been expected of her. The murders and other outrages are forcibly condemned, but then there is an exculpation of the societies themselves. It is only in relation to land that such associations are directed and 'very frequently their proceedings are accompanied by ... sterling traits of unselfishness, generosity, honesty and justice ... In fact, the natural "goodness"... of the Irish peasant is never altogether obscured; and his worst crimes often verge upon the best virtues'![1] This statement is astonishing, coming as it does, from two people who might be expected to be prejudiced in favour of civil law and order. Irish hospitality has always been proverbial, and Mrs Hall is happy to repeat what she has often said – that the stranger will be made welcome in every Irish home, be it grand or be it humble. She recounts many examples of kindness and generosity, she encountered in Irish cottages – 'hospitality without display'.

Contrary to her practice in the earlier works Mrs Hall is at great pains to show Irish people in a better light than their English counterparts. Irish waiters and cab-drivers may be more reckless than English ones, but they display greater warmth and friendliness to their customers. Irish peasants have a natural dignity and courtesy that is not often met with in England, and they are reluctant to take any money in exchange for their freely-offered hospitality. Speaking of a poor family in a wretched cabin who had offered the Halls fresh warm milk, Mrs Hall notes that the woman of the house 'refused, with a half-indignant air, the money we tendered as remuneration'. She adds 'In England we never find any difficulty in prevailing upon this class of persons to accept a silver token of thanks, but in Ireland ... they invariably refuse ... and we have generally found it necessary, when we had given trouble to, or incurred an obligation from a peasant, to present our donation to one of the children, as the only way to avoid hurting sensitive feelings'.[2] Even the ubiquitous Irish beggars, whose existence Mrs Hall has always deplored, are now found to be superior to the English variety. The English pauper is bowed down by his misery, and grumbles and complains without ceasing, while the Irish beggar manages from time to time to make a jest, a quip or a witty remark. English poverty is sullen, that of Ireland is garrulous. Their reaction to alms is very different: 'the Englishman takes relief as a right; the Irishman accepts it as a boon. You may aid half a dozen English paupers without receiving thanks; you cannot relieve an Irish beggar without being paid in blessings'.[3] In her earlier stories and sketches Mrs Hall made frequent references to the squalor in which many Irish peasants lived, and she rebuked them, directly or indirectly, for failing to improve their domestic conditions. In the later work there is a significant change of tone. There are vivid and disturbing descriptions of the dwelling places in which wretched families spend their lives and these are worth quoting:

Not only no window, but no chimney, the chinks in the door alone supplying air and light. The thatched roof is rarely kept in repair, and it is not uncommon for the rain to drip through it, so that one half of its small space is continually in a "sop". Many of them – indeed a majority of them – consist of but one apartment, in which the whole family of grown-up young men and women eat and sleep; there is generally a truckle bed in the corner for the owner, or the "old people" – a cabin will seldom be found in which there is neither grandfather nor grandmother, and affectionate zeal usually cares first for them; but the other members of the household commonly rest upon straw or heather, laid on the floor, covered with a blanket, if it be in possession, and the wearing apparel of the several sleepers.[4]

The pig – the 'gentleman who pays the rent' – and cow, if there is one, and hens and chickens, share the same room at night. Furniture is sparse; the humble bed, an iron pot for the potatoes, a few three-legged stools, sometimes a table, a couple of stone seats on either side of the turf fire, occasionally a side dresser, and a wickerwork potato basket thrown down by the wall. Yet even this poor accommodation is superior to some of the other places the Halls have seen. One that they examined and sketched measured 'ten feet long by seven feet broad, and five feet high, built on the edge of a turf bog'. Seven people lived in there, and their only bed was a 'raised embankment of dried turf', their only covering, in addition to their clothing, a 'solitary ragged blanket'. The Halls emphasised that this squalor was not unusual in Ireland, and, lest they be accused of exaggeration or of citing isolated instances, quoted from the Report of the Select Parliamentary Committee of 1823, appointed to inquire into the condition of the Poor of Ireland, in support of their testimony. The Report found the 'condition of the peasantry to be wretched and calamitous in the greatest degree', and although matters had improved somewhat by the time the Halls made their trips in 1841, there was still widespread and well-attested misery.

The younger Mrs Hall, when writing of peasants living in dirty and degrading surroundings, almost always managed to blame the people themselves for their plight – it was generally due to laziness, intemperance, or lack of forethought. This time, however, she has given the matter more thought, and has seen below the surface. There is a reference to the 'exceeding and unaccountable apathy' of the peasant who lives in such conditions, but who can afford to do better, but then, significantly, there is a paragraph which does, in fact, account for this 'apathy':

This evil will vanish before an improved order of things. It has grown out of long suspicion – a belief that the acquisition of money was sure to bring an increase of rent; a belief not ill-founded in old times. We have ourselves known instances where the purchase of a single piece of furniture, or the bare indication of thrift and decent habits, was a certain notice to the landlord that it was his time to distrain for arrears due; arrears being *always* due under the ancient system when the land was let at a nominal rent – the real value and something above to be paid, and the remainder to be entered as a debt, that kept the tenant in the condition of a slave, utterly and at all times in the power of his master.[5]

Blame is now being shifted from the tenant to the landlord, and Mrs Hall has come out of the mists of memory into the clear light of modern day. The results of this journey into adulthood had already been visible in the sketches that comprised most of *Lights and Shadows of Irish Life*, and were now becoming even more obvious. There had been a regression in the *Stories of the Irish Peasantry*, when she fashioned many of her tales as rebukes to 'her' countrymen and women, but that was now behind her. In one of the stories in that book Mrs Hall had been scathing about those slatterns who kept a dung heap outside the cottage door – indeed the steaming dung heap was, for her, a symbol of dirty Irish domesticity, for it had appeared in earlier sketches – but now in *Halls' Ireland* she has a different reaction. She still shudders at the sight of the dung heap which 'is invariably found close to every door – sometimes, indeed frequently, right across the entrance, so that a few stepping stones are placed to pass over it. It is an evil which is still perpetuated in spite of all appeals on the grounds of decency and health'. Strong words, but she goes on to explain that this 'evil' is unavoidable, for without the manure the food cannot be grown. There is nowhere else to place the essential heap, because 'the cottage cannot trench upon the road – in almost all instances cottages are built lining either a high-road or a bye-road – and he cannot share out of his poor modicum of earth the space thus occupied; every inch must produce its potato'.[6] Mrs Hall's sympathy had always been with the peasants, even though they irritated her, but now she has reached a deeper understanding of the forces that govern their lives and that are revealed in simple domestic arrangements.

In *Lights and Shadows* Mrs Hall had voiced her fears that industrialists would not endanger their capital by setting up factories in Ireland because of reports of the unsettled state of the country and she had assured her readers that Ireland was truly a very safe place in which to live and work. Now, in the work whose chief aim was to generate intelligent and informed interest in Ireland, the assurances of safety were repeated and emphasised, and capitalists in particular were advised that investment in Ireland was not at risk from a disaffected and violent peasantry. Factories in both the north and south of Ireland were flourishing and making a profit for their owners, and the workforce was happy, obedient, and, above all diligent. Mindful of Mrs Hall's earlier strictures on

Irish indolence, a defence was this time made of what might seem to be native laziness. 'It is slander', said the authors, 'to characterise the Irish peasant as an idler, he is often idle, it is true, but it is only because, as often, his time is worth so little as to seem scarcely worthy of consideration.' There is no paid employment to reward his efforts, so the time he spends in idleness is not depriving him or his family of a single 'comfort or enjoyment – much less a luxury'.[7]

In her earlier books Mrs Hall had unfailingly lauded the virtues of Irish women – particularly their modesty, purity and fidelity – and she does so again in the new work. She repeats her criticism of early and improvident marriages, but adds that even poverty never led to estrangement. 'The fidelity of the poor Irish wife', she said, 'is proverbial; she will endure labour, hunger, and even ill-usage, to an almost incredible extent, rather than break the marriage vow.'[8] Irish women are praised as the 'most faithful, most pure; the best mothers; the best children; the best wives'. They are frank and open in their manner, but never coarse or boisterous; they are well-informed but not overbearing; they never 'assume an ungraceful or unbecoming independence', but remain essentially and emphatically feminine'. These qualities are to be found in all classes of Irishwomen, 'from the most elevated to the most humble' and in whatever sphere they move they 'make the best companions, the safest counsellors, [and] the truest friends'.[9] This litany of Irish women's virtues was not composed by Mrs Hall, but, according to a footnote to the text; it was the work of 'that author of the book who was an Englishman'. A graceful tribute from a husband to a wife, to be sure, and one that speaks of a happy partnership. It is interesting that Mr Hall should describe himself as 'an Englishman' while Mrs Hall was ever an Irishwoman (even though there were times, as we have seen, when some of her remarks showed an ambivalence about her nationality). Of the two, both of whom were born in Ireland, it was Mr Hall who had spent longer there, having reached the age of twenty-one before he left Cork, while Mrs Hall was only fifteen when she left Wexford. Possibly Mr Hall's childhood as one of twelve children whose father's business had failed and whose mother was forced to keep a shop, had left him without nostalgic yearnings for the land of his birth, in contrast to Mrs Hall who looked back on a happy and privileged upbringing.

False pride was one of the besetting Irish sins of which Mrs Hall had complained – especially the pride based on the accident of birth. It led, she had pointed out in her previous books, to laziness and dissipation, and in some cases, to a life of crime and a shameful death. She, with Mr Hall, returned to the subject in *Halls' Ireland*

and the earlier strictures were repeated. 'There is', they said, 'a numerous class – almost peculiar to Ireland – of young men possessing the means of barely living without labour; disdaining the notion of "turning to trade"; unable to acquire professions, and ill-suited to adorn them if obtained; content to drag on existence in a state of miserable and degrading dependence, doing nothing – literally "too proud to work, but not ashamed to beg."'[10] This type of young man is to be found in all walks of life and is a disgrace to the country. In general, however, Irish society is healthy, especially among the 'members of the learned professions, and persons on a par with them', and there one can find good humour, witty conversation and 'a universal aptness for enjoyment'. Among Irish peasants, a quality greatly admired by Mrs Hall was the fidelity shown by a servant to a master, and she had written several stories that demonstrated this touching devotion. In *Halls' Ireland* she emphasised this native virtue yet again and related anecdotes of how servants remained faithful through good times and bad, giving their last penny, or their last ounce of strength to help the 'masther' or the 'misthress' when trouble came upon them. She repeated too, her advice to employers, first given in the sketch 'Servants' in *Lights and Shadows of Irish Life*, on how to treat domestic staff in such a way as to ensure their devotion. She put great emphasis on the fact that patience, perseverance and understanding were essential when dealing with Irish servants who were invariably taken from the 'very lowest and poorest in the country', because false Irish pride (and bad treatment from employers), prevented the 'more respectable artisans and peasants from sending their children to service'. Mrs Hall is very much the superior lady, trained in English ways, who, for all her undoubted sympathy with, and admiration for Irish peasant women, is exasperated by their native ways.

Although Mrs Hall's personal views are reflected in the various comments made on Irish life and character in *Halls' Ireland*, the style in which they were written is not always her own, as we have known it in her earlier work. There *are* phrases and expressions which she has used earlier, but there are also whole paragraphs which do not have the sound of her voice. For that we must turn to what is indubitably the lighter part of the book – the legends, the stories, the reminiscences. Most of the recorded memories we have read before, and John Williams, the Bannow postman, Master Ben the school-master, Paddy Cahill the Bannow boatman, Kelly the Piper and Old Frank the coachman, are all familiar figures from the *Sketches of Irish Character*. In spite of this, Mrs Hall manages to invest her pictures with new life – her prose is as fresh and engaging as ever. Some new characters appear – Barney the Natural, Roving Jimmy, Peter Purcel, and 'Pat the Oyster' – and their stories are told with ease and enthusiasm. When Mrs Hall uses her gift for recording Irish speech the words flow effortlessly; it is when she is consciously writing 'proper' English, that she becomes constricted and laboured. There are numerous stories throughout the three volumes of *Halls' Ireland*, consisting mostly of reported speech, poetic and rich in simile. A huntsman 'tattered over the acres like a hail storm'; a youth resembled 'a young eagle in the sun'; little shining waves 'looked like crystal for clearness and yet were as blue as the heavens above'; an outlaw is hunted by the soldiers 'as a dog sets a bird in a field of stubble'; an ailing girl lay on her little bed 'like a bruised water-lily'; a woman's false hair was 'like the fringe on a lobster's leg'; a love-struck young man was 'as blind as a starfish'; and the breath of a dying woman was 'cold as the first breath of the new frost upon the air in harvest'; while a voice murmured 'like a south breeze in a pink shell'. The exuberance and vigour of the language is expressed in original phrases; the oats were left 'to be shelled by the four winds of heaven', the 'rain gathers in oceans' overhead; a gauger was 'made dance the sailor's hornpipe on a hot griddle'; the night was 'as dark a night, Lord save us, as ever fell out of the heavens'; a wife who had 'ever been like a willow in her husband's hand' saw his heart 'turning to iron'; a farmer's wife, chasing away a wicked beggarwoman from her door, said that 'where there's no welcome, the potato has a black heart, and the water's poisoned'; 'all pleasure in sight' left a woman's eyes as her husband walked away from her; and a father whose child had died 'slept the dead sleep of sorrow'.

Many of the stories are legends, some, such as the stories told of the O'Donoghue of Killarney, are already familiar to English readers, for they had 'been frequently referred to by writers who have visited the lakes'. A full ten pages in the section on Kerry are devoted to a spirited re-telling of the O'Donoghue legends, some of it in reported Irish speech, and some in Mrs Hall's own words, simply and attractively told, without undue whimsy. Other familiar legends are those associated with St Kevin at Glendalough, in County Wicklow; these are given in the words of the guide, and, as in her earlier sketches Mrs Hall gives the Irish native pronunciation – 'asy' for 'easy', 'flure' for 'floor', 'haythen' for 'heathen', 'wather' for 'water', 'dhraw' for 'draw', etc. Less familiar legends are also told; the magic sheep whose fine wool made a family rich, the fairy horse known as the Phooka, and the sad story of the seal woman who married a mortal.

At one time Mrs Hall had expressed a certain scorn for the Irish peasant's attachment to superstition, and had scoffed at his simple belief in fairies, or the 'good folk', but in Volume Three of *Halls' Ireland* in the chapter on Donegal there are over twenty pages devoted to Irish beliefs and superstitions. The reason for allotting so much space and attention to the subject is, according to the authors, that the race of fairies is 'daily losing its repute – education and Father Mathew having worked sad havoc among them'. The peasants don't talk so much about them any more, and 'have become, for the most part, even sceptical concerning them; and deliver their anecdotes with an air of doubt at the least, which indicates an abandonment of their cause approaching to contempt of their power'. Reason and common-sense are taking over from the imagination, and before long 'the Irish peasant will retain little or nothing of a distinctive character'. There is a distinct note of regret in these comments, but although 'the mere searcher after amusement' may regard the change as a misfortune, for those 'who have higher hopes and objects, the change provides a theme for grateful rejoicing, as inevitably tending to incalculable good'.[11] This prim assertion is followed by copious quotations from Thomas Crofton Croker's book on Fairies, *Fairy Legends and Traditions of the South of Ireland*, and then by some stories of Mrs Hall's own – or at least, told in her style, using reported speech, and being lively, well-rounded narratives. The more one reads of the composite production, the more the contrast between the stories of Mrs Hall and the statistics of her husband jar upon one. The change of mood and tone is so abrupt as to produce a feeling of anti-climax, almost bathetic at times.

The big question of course, is how much of the work can be attributed to Mrs Hall, apart from the stories, legends and personal reminiscences. Some of the pompous prose has already been seen in her earlier work, but what if that had been the work of Mr Hall? It is not inconceivable – he was a professional journalist, an expert, an editor, and she was the amateur, a woman without any experience in the world of writing. We know, because she has told us herself,[12] that she read out her early efforts to him for his guidance and approval, and it is possible that he made emendations and corrections and suggested how the work could be improved, or made acceptable to editors. There is no proof that this is what happened, but the possibility is strengthened when we read two letters written after Mr Hall's death in 1889 and published in the

Athenaeum, The first, on March 23rd, 1889, is from a Mr Purnell, a journalist and friend of the Halls, who, referring to the obituary of Samuel Carter Hall that had recently appeared in the periodical, states that he 'could say from knowledge that the husband was the guide and counsellor even in the wife's charming tales and novels'.[13] On April 6th another letter appeared in the *Athenaeum*, 'as a sort of corollary'. This one came from Emily M. Hickey, grand-daughter of the Reverend William Hickey ("Martin Doyle", the writer on agricultural affairs), who, with his wife, was an intimate friend of Mr and Mrs Hall. She quotes from a letter to her written by Mrs Hall in 1872, returning some poems which Miss Hickey had sent to Mr Hall for his criticism:

After giving me her husband's criticism on some of my poems, Mrs Hall went on: "You know I never write poetry, but often, often, Mr Hall, when going through one of my tales, has said, 'My dear, so-and-so is the case. You have given words, instead of thoughts. Destroy this page, *think*, and re-write it.' And such is my faith in him, that I never *disputed* his judgment, but *did* as I was bid. And frequently the monster would make me do it over and over again until it became what I saw was right."[14]

This is not proof positive of interference on Mr Hall's part in the writing composition of Mrs Hall's stories, but it certainly increases the likelihood of his having influenced her writing. From what we read of Mr Hall's own undoubted work in the *Art Journal* and elsewhere it is clear that he was lacking in any great literary ability, and was given to pomposity and sententiousness. A contemporary, the publisher Henry Vizetelly, wrote dismissively of Samuel Carter Hall in his autobiography, published in 1893. According to Vizetelly, Hall was the original of Dickens's Pecksniff, and was 'devoid of the slightest critical faculty, possessing only commonplace taste in matters of art, without even the power of expressing himself logically'. Vizetelly referred to Mrs Hall as being 'far more celebrated' and added: 'Hall talked even far more priggishly and foolishly than he wrote, and I have more than once felt surprised at hearing him launch out at his own dinner-table without any attempt being made to check him by his sensible wife. True, he assumed an intellectual superiority over her, and she blandly accepted the false position, but no one was taken in by this'.[15]

Mrs Hall, for all the sweetness and sentimentality of her fiction, was a realist, and if 'blandly accepting a false position' was what was needed for domestic comfort and happiness, then so be it. In a story, 'The Mosspits', published in the *Amulet* of 1832 she had commented:

It is a severe trial of a woman's judgment if she discovers her mental superiority to the lord of her affections, and yet, while she secretly manages all things for the best, makes the world believe she is only the instrument of his will. A wise woman *will* do this, it is only a wise women who *can*.[16]

If she was referring to her own marriage then Mrs Hall was indeed a 'wise woman', for the Halls, although they had no family, built up a good partnership over the years. There is no reason to believe that they had anything but a stable and happy marriage and it is probable that he influenced her writing for the worse, weighing down her simple and attractive stories with ponderous reflections and moral maxims. Still, there must have been something in her own character that welcomed these additions, if such they were, and believed that they added lustre to her work. In any case, moralising was an essential constituent of the fiction of the time, and it would have been very unusual for a writer of Mrs Hall's calibre not to have adhered to the norm. She was not an original or revolutionary thinker so there would have been no conflict between her and her husband, either in life or in work.

The inclusion of the reminiscences, stories and legends, in the main text of *Hall's Ireland* is clumsily done at times. They do not flow easily or logically from what the reader is being told about local beauty spots, natural wonders or places of archaeological interest, but are baldly set down in an obvious attempt to sustain reader interest. The authors themselves are aware of this and ocasionally make semi-apologetic introductions. For instance, having devoted five pages to the subject of Irish music, and having described in detail the various instruments which were used, they say 'As the treatment of this subject, however necessary, may appear dull and heavy to the general reader, we ask leave to introduce a sketch of an old piper',[17] and what follows is Mrs Hall's reminiscence of 'Kelly the Piper', whom she knew in her youth. Elsewhere, the thoughtful essay on the evils of the land system in Ireland and the consequent rise in violence, is followed by a story about 'an industrious and respectable young farmer' who is lured into a secret society. This is prefaced by the explanatory statement: 'In pursuance of our plan of illustrating the leading characteristics of Ireland by the introduction of "a story", we entreat the attention of our readers to the following – premising that it is but a very slight colouring of a circumstance that actually occurred within our own knowledge'.[18] Elsewhere they wished 'to relieve the monotony' of their archaeological details by a story, or 'to lighten the heaviness of the preceding pages' by telling tales about interesting Irish characters – process-servers, wise-women, beggars or local idiots.

Although most of the reviewers of *Halls' Ireland* were happy with Mrs Hall's stories and anecdotes there were occasional dissenting voices. The writer in *Fraser's Magazine* was highly critical of what he called her 'sentimental nouvellettes'. In his review of the first part of the work he opined that 'these stories were out of place in a book of which the general narrative professes to give the actual facts which came under the observation, or are the result of the inquiries of the author, in the ordinary narrative style of tourists'. The description of beggars was 'graphic and amusing because ... people must be convinced that they are true'. But the sad stories that Mrs Hall reported as having been told by two unfortunate women were so obviously fictional as to be out of place. No peasant girl or woman could speak so tenderly or so eloquently and consequently the stories 'disturbed the verisimilitude of the whole'.[19] A similar criticism was voiced by the reviewer of Part I in the *Literary Gazette*. Although 'taking it for granted that the reputation of Mrs Hall in the treatment of Irish subjects may be deemed a guarantee for much to interest', he found in the work 'a little too much of the ornate which belongs to fiction, for this is to be a real and actual picture of Ireland, to inform the British people', and he cautioned Mrs Hall about telling 'heart-appealing stories'.[20] A contrary opinion was voiced by the critic in the *Dublin University Magazine* in his review of Part I of *Halls' Ireland*; he greatly admired the tales which were 'told in Mrs Hall's best style', and the whole production was 'so free from dryness and insipidity, as not to produce tedium in the most devoted admirer of "light reading", and yet with sufficient solidity to satisfy the most strenuous advocate of seriousness'.[21] The *Atlas*, reviewing Parts 1-5 of *Halls' Ireland* singled out for special praise the portions illustrating the character of the Irish people. It was 'drawn out ... with force and truth', but this was only to be 'expected from the previous writings of one of the authors'. There was no need to mention Mrs Hall by name, all who were familiar with her writings would know her skills, and the extract quoted from the work under review was obviously from her pen.[22] *The Sunday Times* in its review of the complete work emphasised the importance of Mrs Hall's contributions. According to this critic she

> infused soul into this production. There is a gushing tenderness, blended with power, in all Mrs Hall writes. Her strong sense of indignation at unprovoked tyranny shields all she produces from the charge of effeminacy, and yet (Heaven be thanked!) she is never masculine ... Beyond all question Mrs Hall is the most simple, yet energetic, powerful and truthful and pathetic writer of the day.[23]

The Irish Nationalist periodical the *Citizen* took a different view of Mrs Hall. In a review entitled 'Mock-Irish Works', which mused in a general way on guide-books and their misleading accounts of countries, the critic went on to sigh over what he saw as the caricature of Ireland presented by the Halls in their work. This production was 'anti-national', and although he hadn't expected much in the way of praise of Ireland from the Halls he had not envisaged 'elaborate, serious and systematic depreciation'. A direct criticism of Mrs Hall followed:

The clever, though perpetually half-Irish sketches of Irish life, for which the English public have long been indebted to Mrs Hall, were not with us exciting topics of either praise or blame. We took it for granted that the worthy authoress really believed they were true likenesses of character and society here; and as we saw no likelihood of any great harm arising from the errors they contained, we chose rather to acquiesce than to quarrel.

But whether we are to attribute any or all of the increased disposition to burlesque and disparage her native land which her new work displays, to the circumstance of her husband having united in its comportion or to some other cause to us unknown ... we feel that it were a tame surrender of the rights of popular censorship, if we suffered such a work to pass ... without recording our marked and almost unqualified reprehension of its tone.[24]

Praise or criticism are directed more towards Mrs Hall than to her husband, a clear sign that she was recognised as the more established and more accomplished writer of the two. Reviews of *Halls' Ireland* that appeared in English and Scottish provincial papers (and were reproduced in a publisher's advertisement in the Fifth Edition of *Sketches of Irish Character*) speak of 'Mrs Hall's reputation in the treatment of Irish character; her 'great literary attainments', even 'her genius'. By contrast Mr Hall's is 'a graver pen', he is 'a convincing and elaborate essayist', and 'a man of talent'. This was decidedly not the view held by the Irish periodical, the *Nation* which stated

Mr Samuel Hall is a Cockney who married a Wexford woman and spent a summer or two in Ireland. He and his wife issued a catchpenny book on this country, made up of gossip and plunder. Some of this plunder was good enough. Its sale in Ireland was trivial (600 copies); in England it reached 8,000, as we have heard. The Irish knew it to be weak and conceited – the English were humbugged into believing it clever and fair.[25]

Here the criticism is directed at both authors, as is that in the *Dublin Monitor*, the fiercely Evangelical weekly, which reviewing Lady Chatterton's *Irish Sketches* said that 'her descriptions of Irish scenery are worth a cart-load of the "Hall humbugs" with their gross caricature and exaggerated commonplaces'.[26] The *Monitor* was not without bias – it was of the same religious persuasion as the proselytising Reverend Edward Nangle – whose mission on Achill Island had been severely criticised by the Halls in their chapter on Mayo. A furious correspondence between Hall and Nangle followed in the *Monitor* in the spring of 1841, without resolving anything, or moving either man from his entrenched position.

Religious susceptibilities of the day were so tender that it was those portions of *Halls' Ireland* that aroused the fiercest criticism. The Halls anticipated this, saying, in reference to the controversy about the college at Maynooth, 'Let no one consider our remarks upon this all-important subject out of place. To have written a book concerning Ireland, and to have passed over the source in which so vast a portion of its prosperity or misery must originate, would have been an omission for which we could have urged no satisfactory excuse'. Their own solution to the problem of national religious education would be 'to *exclude all direct religious education from the schools*, and to intrust that most essential part of the training of youth to the pastors and teachers of the pupils, either at their own homes, or in their own places of worship'.[27] This suggestion would have been contrary to all that the Catholic *Dublin Review* stood for, and probably accounted for the bitter tone of the critic's assessment of the complete edition of *Halls' Ireland*. He called it 'an interesting and most insidious, and therefore, most dangerous work'. In their 'libel upon Maynooth College' the authors made no less than 'five gross and calumnious falsehoods, in matters of plain fact, in the short space of *four* lines! Truly', said he 'this *is* the age of reckless slander'.[28] If Mr Hall was the gatherer of facts and statistics, as he was generally supposed to be, then this criticism was directed at him, and not at his wife, but, as they shared the praise, no doubt they also shared the blame. Their partnership was a great success in terms of sales, (in spite of what the *Nation* had said) for *Halls' Ireland* was followed by a series of handbooks on individual counties of Ireland (essentially the same as the major work, but with some slight differences), and was itself translated into German in 1850: *Skizzen Aus Irland*. A new edition of *Halls' Ireland* was published by George Virtue in London in 1853 and contained some additional material to bring the original up to date. The census returns for 1841 were included in this edition, as was a supplementary paragraph describing the visit of Queen Victoria to Dublin in 1849, when 'all were gratified with Her Majesty's affable and condescending manner'. In the new edition, the original account of the working of the Poor Law was followed by a short paragraph adverting to the ravages of the famine of 1845 to 1848. 'To meet such a calamity no system of Poor Laws could be framed', the Halls admitted, adding that the 'laws, compulsory rates and voluntary gifts, national loans of millions and private subscriptions ... were alike insufficient'. However, this was a sad subject and one which was unsuited to the 'character' of their work, so, although it was one which could 'not be wholly passed over in silence', the Halls felt bound to spare their readers 'the perusal of such a chronicle of human misery'.[29] The original edition of *Halls' Ireland* is a joyous affair, full of hope and optimism and belief in a golden future for Mrs Hall's 'own dear country'. It is easy, when reading it, to be caught up in the authors' own enthusiasm and to believe, with them, that for every new visitor Ireland received it would obtain a new friend. Famine and fever were still local occurrences in 1843 when the three original volumes appeared, and a tourist could travel in comfort and safety through the country without seeking more than a few examples of Irish peasant poverty. Ten years later it was a very different story, but the Halls evidently felt that their position as tourist guides did not require them to enquire deeply into the causes of the calamitous famine or to describe its harrowing effects.

EIGHT

The Whiteboy –
'A truly national novel'

In his review of Mrs Hall's *Lights and Shadows of Irish Life* in 1838, the writer in the *Dublin University Magazine* expressed the wish that she might one day write a novel set in Ireland – 'a truly national novel'.[1] However, although Mrs Hall wrote four novels with an English background – *The Buccaneer* in 1832, *The Outlaw* in 1835, *Uncle Horace* in 1837 and *Marian* in 1840, it was not until 1845 that the reviewer's wish was granted and a completely Irish novel was written. This was *The Whiteboy; A Story of Ireland in 1822*,[2] and it could fairly be called a national novel, because not only did all the action take place in Ireland, but Mrs Hall, through the medium of the book, set out the problems of the country, confronted them, and offered her own solutions. It was a brave undertaking, and an ambitious one, and the book is of intense interest for the way in which it contains the distillation of all Mrs Hall's views and opinions about Ireland. Her previous writing career was, one can now see, an apprenticeship for this finished work.

The story of *The Whiteboy* is a simple one and one that evokes echoes of Maria Edgeworth, Lady Morgan and Eyre Evans Crowe. Edward Spencer, a young Englishman, inherits an estate in Ireland and travels there to take possession. He knows nothing of the country but is suffused by a romantic love of its music and poetry. After a short time in the country his eyes are opened to reality, and he determines to do what he can to help his new-found countrymen. He marries a local girl, his cousin Ellen, who has been living as a poor relation on his uncle's estate, and settles down on his estate vowing to run it properly and to care for his tenants with justice and with mercy. There are several sub-plots, notably one involving the Whiteboy of the title, which give Mrs Hall the opportunity of commenting on several features of Irish life which needed attention, but the central thesis is the relationship of landlord and tenant. This had been a pre-occupation of Mrs Hall's from the very beginning of her writing career.

It was inevitable in the conditions obtaining in early nineteenth-century Ireland that Mrs Hall should so often use the landlord/tenant relationship as the core of her writings. Millions of words were written about this relationship in contemporary periodicals, travel books, short stories, novels and other publications. It gave ample scope to moralists, economists, agriculturalists, polemicists, jurists and writers of fiction alike. The fact that it was tirelessly and intensively debated shows that there was an awareness that all was far from well in the relationship. The ramifications of the problem are so widespread that it is virtually impossible to deal with them in a small area so we must make do with a limited examination of what was said, and concentrate on the moral problem. There was no debate about the morality of owning land and leasing it out to others, it was how the ownership was exercised, and how the leasing was carried out that was in question.

The patron-client relationship, as the anthropologists call it, was peculiarly important in Ireland, where a colonial mentality assumed stupidity and indolence on the part of the natives, and where greater efforts than usual were required from the landlord if the tenants were to keep their side of the bargain. The first duty of the landlord was to reside among his tenants and give good example. This dictum is repeated over and over again, in farming magazines, in pamphlets and in fiction. A typical example from the farming press is found in the introduction to the first volume of the *Irish Farmer's and Gardener's Magazine*. Speaking of the obstacles to progress in agriculture the writer says:

Absence from the country of so many of its landed proprietors whose example of improved cultivation were they to reside on their properties, and to devote, or cause to be devoted, the requisite degree of attention to this subject would exert a salutary influence on the district favoured with their presence, must also be mentioned as an obstacle to its improvement.[3]

A mild comment, free of any emotionalism, but to the point. It was not enough to be a resident, one had to give good example and one had to develop the land one owned. Unfortunately, some residents failed in this duty, and in the same volume of the magazine a writer points out the financial advantage it would be to them if they did fulfil their moral obligation:

Would our landlords attend to the real improvement of their tenantry, by affording them the means of improving their farms, and endeavour to enforce improved methods of husbandry – for it will take force to accomplish the object in view – their own rent-roll would soon feel the beneficial effects of such a practice, and the altered condition of the now miserable and half-fed population would surely be a matter of gratification to every lover of his country.[4]

Mary Leadbeater, the Quaker from Co. Carlow, who took a personal interest in the problems of the peasants in her lively little works of practical and moral instruction, expressed very clearly this belief that it was the duty of an Irish landlord to work for the betterment of his tenants, not only as an inherent obligation but as an act of charity because they were so greatly in need of help and direction. In *Cottage Dialogues among the Irish Peasantry* (1813) she has Barney, the tenant, seek advice on renting land from his landlord, Mr Seymour. He is told what rent would be economic to pay, how best to cultivate the land, and is encouraged to seek advice at any time. Thanking Mr Seymour, Barney says:

"Gentlemen have no notion what good they might do to people, by considering for them, and advising them; for gentlemen have such opportunities of seeing how work is done in different places, and of reading in books about the nature of land, and corn and cattle, that when they give their mind to the like, and live among their tenants, and put them on good methods, they are a blessing to the country."[5]

This instructive passage is not, of course, directed at the Irish peasant who might possibly be reading it, or having it read to him, but to the Irish landlord and member of Mary Leadbeater's own class, who needed reminding of his obligations. In a review of a new edition of *Cottage Dialogues* in 1841, the critic writing in the *Citizen or Dublin Monthly Magazine* pointed out that Mary Leadbeater gave an example of 'a good landlord – a rich man *that resides at home* to take interest in his tenants and advise them: a lady that lives among her people, instead of at a foreign court, and excites them to honest industry by her interest and superintendence'.[6]

There is an admirable subtlety in Mary Leadbeater's portrayal of the peasants' perception of landlord bounty. They are truly grateful and are obviously worthy recipients of good advice and interest, so that the class of reader whom the author is addressing will feel it worth his and her while to continue in their good works, which is, after all, the main point of the books, but they voice the discontents which the landlords may sense among them. Critical opinions are put forward by one peasant speaker but are then swept away by another. For instance, in Part I of *Cottage Dialogues* a young woman, Mary, who has been taught to plait straw for bonnets by 'a very good young lady' who did all she could 'to help poor people by teaching and giving them work to do', expressed her thanks and admiration: "Good luck to the quality, though we are apt to think they don't care about us, yet when we consider all the schools they set a-going, and all the industrious little things they put us in the way of doing, we feel they are our best friends". There is a thread of peasant doubt there: 'apt to think they don't care about us', but her friend, Rose, brushes it away with her reply: "I am glad to hear you say that, for poor people often give themselves the say of scoffing at their grand neighbours, and grudging them their fine houses and clothes, while those very neighbours keep bread in their's [sic] and their children's mouths".[7]

In Part 2 of the *Dialogues* the same theme is taken up by the peasants, Thady and Martin, who are speaking of local hospitals. Martin believes that such places must cost a great deal of money and that "poor people ought to be very thankful, that the quality are so good as to keep up such places for us". Thady looks at it differently: "What signifies what you or I would call a power of money to the quality? They have so much, they don't know what to do with it". Martin is outraged and upbraids him for his foolish talk:

"Look about you, and see the grand houses and the beautiful gardens that they have, and don't you think it takes a great deal of money to keep them up? and how many people are made the better by working at them; and then the company that comes to see the quality, and the journeys they take to see them again; besides the horses and carriages, and all the other things that they want, that we have no notion of, run away with a sight of money".[8]

The author is delicately criticising landlords through the person of a sceptical peasant, and more tellingly, through the defence of them made by a loyal tenant, humbly accepting that he lived in a world more drab than that of his landlord and the landlord's friends.

'Martin Doyle' (Reverend William Hickey) in his *Hints to the Small Holders and Peasantry of Ireland* (1830) had advised his peasant readers to be respectful to the gentry, because they would then be 'more likely to live among you and promote your welfare',[9] and in his next work, *An Address to the Landlords of Ireland on Subjects connected with the Melioration of the Lower Classes*, (1831) he advised the landlords on how to behave. He listed the reasons for the existing misery of the peasants – want of employment, the 'indolence and inattention of a large portion of the Irish proprietary', their want of enterprise and their 'neglect of the minute and personal superintendence of their estates'. To these he added absenteeism and the evictions following the abolition of the forty-shilling freehold. All these problems had a solution. If the landlords were more enterprising and made better use of their land, more people could be employed. If, however, in spite of all improvements – drainage, hedging, land reclamation, road making, and so forth – there were still too many tenants, then a scheme of assisted emigration should be provided. Money would be needed for all this but if landlords curbed their extravagances then it would be available and the 'superfluities of one class should administer to the necessities of the other'.

His central message, though, is a moral one. The landlord has a duty towards his tenants and if he fails to carry it out then he must expect to suffer the consequences. His kindness, his interest, his acts of charity, and above all, his good example are what is needed. He is straightforward in his allocation of blame:

When we hear of districts filled with resident gentry, in which crime prevails to an alarming extent, and where the lower orders are uncivilized, it is fair to conclude, that, in such cases, the local proprietary is unfaithful to its trust, that *personal inspection does not take place*, and that consequently the humanizing influence of honourable example, of kind and conciliating manners, and of affectionate solicitude, is unknown, as its power is unfelt. In such places it is to be inferred that the local gentry are, in a great degree, themselves to blame.[10]

Martin Doyle examines in more detail the ideal relationship between landlord and tenant when he comes to deal with absenteeism, 'which has been regarded as a leading evil by every practical man who has witnessed its mischievous effects'. He accepts that there are instances where absenteeism is unavoidable and where a good agent may act as substitute for the landlord, but in general it is to be deplored because of the bond which should exist between the tenant and his landlord. He laments the 'squalid negligence, the lounging laziness, and despairing aspect of a tenantry without a landlord to whom they might apply for counsel, encouragement, indulgence or relief'. The peasantry are nothing without their landlord's good example, they can do virtually nothing for themselves and unless he remains in close communion with them the whole system will fall apart: 'By this unnatural separation [absenteeism] the social duties are abandoned and their connecting links, on which depends the very existence of civil life, are broken asunder'.

Martin Doyle's insistence that there is 'a chain of indissoluble interests' connecting landlord and tenant was often repeated over the years with emphasis on the 'friendship' that was supposed to exist between them. There are references in the farming papers to the vigilance (as gamekeepers) of the 'affectionate tenantry', 'the bonds of reciprocal kind offices', the 'community of interests', and of special interest as revealing what a good landlord thought his duties were is the address delivered by 'that excellent resident gentleman', Robert Fowler Esq., to the tenantry on the manor of Moylough, and reported in the *Irish Farmer's and Gardener's Magazine* (February 1838). He said that

"By an improved system of farming, the tenant becomes more industrious, improved and cleanly in his habits – puts more money into his pocket, and keeps his land in better heart – which advantages certainly are of benefit to the landlord, by his farm's being more valuable, and the tenant more punctual in his rent. There are different ways of working – and the landlord can compel his tenantry ... to crop their land as he likes; but I wish to induce you to do so on the Christian principles of confidence and affection, instead of compulsion. Believe me, my friends, your landlord is your best friend and where he resides on his property, should be more frequently consulted than he is – and my sincere desire is to benefit you all."[11]

This vision of the landlord as sincere friend and mentor of the tenant found its way into fiction and from Maria Edgeworth on he was portrayed as the one person in authority whom the tenant could truly trust. The belief that it was he who could protect his peasantry from rascally agents and middlemen was nurtured by many affecting stories of his intervention, and up to 1845 he was very often shown as the tenant's saviour, as well as the model on whom the tenant should pattern his behaviour. William Carleton, a peasant writer, with no love for landlords, nevertheless recounted in his *Traits and Stories of the Irish Peasantry* (1830) several instances of good relations between landlords and tenants. In 'Shane Fadh's Wedding' the Squire and his son attend the party and are greeted with enthusiasm by the whole company, and the tie between the landlord and tenant is emphasised in 'Larry M'Farland's Wake' when it is recalled that the landlord's family always attended the M'Farland children's christening and even stood as god-parents. It is in 'The Geography of an Irish Oath', however, that Carleton most clearly depicts a relationship between a landlord and his tenants as one of mutual respect and sound good sense. The landlord is no absentee but lives on his estates and takes a keen interest in the welfare of his tenants. He 'instructed them in the best modes of improving their farms', and was sensible enough not to demand from them rents that were too high for them to pay. In return he expected that his tenants should work hard and improve their holdings. As Barbara Hayley has pointed out, this sketch prevents

The Whiteboy – 'A truly national novel'   151

'Carleton from appearing to advocate tenant right at the expense of the landlords. In it he demonstrates that the country and its farms can prosper, but only when landlord and tenant are equally thoughtful and businesslike'.[12] This landlord's involvement with his tenants, the Connells, goes beyond a simple business arrangement, and becomes a more personal affair. The Connells are listened to with interest and with patience and when Peter Connell begins to drink heavily after the death of his wife Ellish, the landlord tries to deflect him from the path of self-destruction, saying: '"I have uniformly been your friend, and the friend of your children and family, but more especially of your late, excellent and exemplary wife"'.[13] Mr Herriott, one of Mrs Hall's favourite characters, was described in 'Kelly the Piper' in *Sketches of Irish Character* (1829), as 'having a kind and benevolent temper; he loved to see the peasantry happy in their own way, and spent his fortune on his estate; anxious both by precept and example, to instruct and serve his tenantry'. His efforts were appreciated and he was accorded every respect when he attended rural gatherings:

The bustling and the skirmishing instantly ceased. The men held their hats in their hands, and the women rose and curtsied respectfully, as Mr Herriott and his family proceeded, while many a heartfelt blessing followed their footsteps.

Perhaps the most delightful prospect in the world is that which a good Irish landlord enjoys when his tenantry are really devoted to his service; because their devotion is manifested by those external signs, which can only emanate from an enthusiastic temperament.[14]

According to the works of fiction in this period, there was a distinction made by tenants not only between resident and absentee landlords but also between those of the old school or class, and the newer ones. Nostalgia paints the past in a rosy light, and the virtues of the older landlords may have been exaggerated by the authors, but the fiction rests on fact – at one time the forty-shilling freeholders had votes which were of value to the landlords. When the franchise was abolished such freeholders had no political value any more and an unscrupulous landlord would do his best to get them off his land. Mrs Hall, writing in the 1820s, was aware of peasant feeling in relation to the new breed, and adverts to it in several of her sketches and stories. Mary MacGoharty, when asking the narrator to write a petition for her to young Squire Bromly, tells a story about the old Squire which does not do him much credit, but mentions in passing the fact that he was once opposed at an election by a man named Jack Johnson and adds: '"Squire Jack, they called him – though I was only a girleen at the time, I never could turn my tongue to say Squire Jack, and he only a bit of a brewer'". Mr Herriott, the Squire featured in 'Kelly the Piper' was an 'oul residenter' and Mr Barry, owner of the house described in 'Hospitality', was praised because he '"had nothing in particular to distinguish him from the rale true-born gintry"'. In 'The Bannow Postman' a young woman, Mrs Clavery, who begged for alms at the Parsonage and was fed and sheltered by the kindly family there (described by the eponymous postman as '"the true breed of the gentry ... Ah, the thick blood without any puddle for ever!"') told the story of how she had been ill-treated by her landlord and his wife. The wife had demanded spinning and knitting as "a yearly compliment" and when it failed to arrive one year because of Mrs Clavery's illness, she took some of the poor woman's best hens. She was a grasping woman '"not a born lady"', observed Mrs Clavery, '"and they're the worse to the poor. Musharoon gentry! that spring up and buy land, hand over head, from the raale sort that are left, in the long run, without cross or coin to bless themselves with; all owing to their generosity"'.[15]

Among the newly-rich landowners were men who had progressed from the position of agent to a landlord to being landlords themselves, in some cases on the very property on which they had once worked. They were known to have given loans to their original employers to help them out of their financial difficulties caused by encumbered estates, inherited debts and the debts incurred by their own extravagances. They were suspected of having cheated both landlord and tenant on their rise to power and were generally disliked and distrusted as landlords. Squire Johnson in 'Mabel O'Neill's Curse' by Mrs Hall, is confronted by the woman he has wronged and is reminded of his lowly origins: '"In those days this was yer employer's house, but he earned his gould, and then he borrowed it and ye lent him back his own – ye may well turn pale, it's all true"'. In the epilogue to the story the hero says of Johnson: '"The higher he got, the more was the finger o'scorn pointed at him, for he hadn't the gentlemanly turn about him"'.[16]

The distinction between landlord and agent is made clear in the fictions and the message is obvious – there are some bad landlords, but the agents are generally much worse. In fact, a favourite fictional image of the landlord is that of the power who will save his tenants from the cruelties of the agent. The landlord's chief fault is negligence rather than malice, and the belief that all will be well if only the landlord might learn the true state of things and waken up to his responsibilities and duties is shown as a widespread one among the tenantry. Kate Connor, in the story of that name by Mrs Hall, makes her way to London to seek the help of her landlord when she and her family have been evicted by his agent, whom she designates first and foremost as being no gentleman, neither "'by birth nor breeding ... sure if ye put a sod o' turf – saving yer presence – in a goold dish, it's only a turf still'". His lordship and his daughter Helen, who narrates the story, travel with Kate back to Ireland, the wrong is righted, the agent dismissed and the landlord resolved "to reside in Ireland for six months out of the year".[17]

When the landlord was an absentee there was no chance of forging those mutual bonds of friendship and esteem that formed the ideal relationship of landlord and tenant; and the writers of fiction, no less than the writers on political economic and agricultural matters, identified absenteeism as a major cause of Ireland's miseries. Mrs Hall, speaking through the character of Alick the Traveller in 'Annie Leslie' says bitterly that the Lord and the Lady "get the cash – that is, as much as the agent chooses to say is their due – and spend it in foreign parts, without thinking of the tears and the blood it costs at home".[18] Sir John Clavis in 'The Last of the Line' was an absentee for a time (and so allowed his bailiff, Denny Dacey, to gain a financial hold over him), and Kate Connor's landlord, as we have seen, had abandoned his estates to the care of an agent.

The greatest crime the absentee landlord could commit was to leave his estate in the hands of a dishonest agent. Not all agents, of course, were dishonest or grasping and cruel, and some served their employers well, seeing their interests as paramount, and being intelligent enough to realise that contented tenants were the basis of a well-run, efficient and prosperous estate. They fully adopted the landlord ethos and a typical example of their beliefs may be found in a speech given by Mr Charles Doyne, agent to Lord Beresford, to the Agricultural Society in Carlow in 1834. As reported in the *Irish Farmer's and Gardener's Gazette*, he said:

It has ever been my object to exert every energy in promoting the prosperity and happiness of the tenantry placed under my care; and in so doing, I consider it the most efficient way of studying the interests of the landlord. There is between both parties an identity of interests which I shall ever regard, and a reciprocity of obligation which should not be forgotten. The landlord and tenant are bound to each other by mutual bonds of friendship and esteem; and if, unhappily, these ties are severed, the tenant practically separates himself from his benefactor and stands alone and unprotected upon his own resources.[19]

However, most agents in fiction were wicked and dishonest. We have seen how tenants appealed to their head landlords for protection against their rapacity, and how the landlords themselves were often their victims. The fictional accounts of agents' cruelty towards the tenants all focussed on their greed, which manifested itself even in petty ways, such as taking eggs and hens from poor peasants and was at its worst when land was in question. Their aim was to get land, the true source of wealth, into their possession and then rent it to the highest bidder. They obtained the land by cheating both the landlord and his tenant, a simple matter when a landlord took no personal interest in management. When a lease expired it was easy for the agent to refuse its renewal until a higher rent was paid, and the landlord knew nothing of this, and so no extra money came into his pocket but went instead into that of the agent. Bribes were also demanded and were a steady source of income. Eventually, the land-hungry tenant, or the man who wanted to cling to his family home, was unable to pay the high rent, and the land fell into the hands of the agent. A man as unscrupulous as this had no qualms about evicting tenants, no matter how pitiable their condition might be, and Kate Connor gives an affecting account of how her mother was turned out of her cottage '"when the snow was on the ground – in the could night, when no one was stirring to say, God save ye"'. Mary Clavery too, in 'The Bannow Postman' had her sad story of injustice: '"T'was a sorrowful parting, for somehow a body gets fond of the bits of trees even, that grow up under their own eye, and I was near my lying-in and the troubles came all at once and all we could get to shelter us was a damp hole of a place"'.[20]

Mrs Hall in *The Whiteboy*, paints a picture of an oppressing agent, Abel Richards, who in a very short time before his employer, old Mr Spencer, found him out, managed to establish himself as a landlord. She was rather vague about how he acquired such power but very clear about how he exercised it. He allowed his tenants to go into arrears and then, as soon as he saw they were in poor circumstances, he called in the debt. He evicted without mercy, and had no pity for the widow or the orphan. He is an overdrawn figure and impossible to take seriously but he typifies the breed. The agents became landlords in their own right as a result of cheating their employers and the peasantry, and the strictures on bad head landlords apply equally to them. Their own wants and ambitions were paramount and they felt no sense of duty towards their tenants. The high-sounding principles enunciated earlier in the century when a duty of care was enjoined on landlords meant nothing in the face of greed and self-interest. Even practical considerations had no force, and by penalising good tenants for the improvements made to property the agents sacrificed long-term prosperity to short-term returns. It is not only with hindsight that this is obvious, it was widely commented upon at the time, especially by agriculturalists. The *Farmer's Gazette and Journal of Practical Horticulture* in 1844 begged landlords 'to contrive by every means in their power, to give an impulse to the improvements thus happily commenced; first, and above all things, by satisfying their tenants that they have a fair prospect of reaping the reward of their industry'. The contrary was the case, and the 'nefarious and monstrous custom of ejecting tenants who have made improvements, or when permitted to remain, making them pay for the improvements which they have made,[21] was the cause of depressing the value of the property, as well as stifling the spirit of the farmers'.

One of the big problems in Ireland facing the head landlords was the sub-letting and the sub-division of land. In the latter case a family divided its holding between the sons and the sons' sons until in time these were small uneconomic strips. The agents sub-let the landlord's land, and their own, and the sub-letting of sub-let land also resulted in minute holdings. Apart from the obviously bad agricultural practice of having quarter or half acres as holdings, there was a social consequence – the dissolution of that bond which should ideally exist between a landlord and his tenant. The fact that it often did not exist even where holdings were substantial and contact between landlord and tenant (or agent and tenant) was real and personal was beside the point, because at least the conditions for its existence remained, and a change of personnel might bring about its flowering. In the other situation, where a landlord was separated from a crowd of miserable wretches by a sequence of 'middle-men, and third-men and fourth-men' there was no chance of a bond growing up. The rents charged by these middle-men were exorbitant, not so much of themselves, as because the holdings were incapable of producing the money to meet them.

There was another problem inherent in the landlord/tenant relationship – the attitude of the tenants. In the few instances where there were good, conscientious, and resident landlords they came up against a certain amount of peasant intransigence. Landowners and others 'who possessed influence with small farmers' were advised by the *Irish Farmer's and Gardener's Magazine* in 1834 'to forward, by all means, the introduction of a proper system of agriculture among them, and to discourage, in every case, the too prevailing disposition to stick to the old and unprofitable usages of times when increased means of supporting human life were not so much wanted'.[22] Mrs Hall had often mocked this indolence and conservatism and in *The Whiteboy* she went into the subject of Irish resistance to new ways in great detail. After some trouble and some difficulty the young English landlord, Edward Spencer, is established on his Irish estates, and with his Irish wife, Ellen, devotes himself to improving the lot of his tenantry. It is not easy because they are still attached to their old slovenly ways, and 'it takes time and patience, great patience!'. It was worth it, however, 'and every small holding in the vicinity partook of the landlord's prosperity'. The cottages increased in number though 'some, to be sure, were not as neat or as well kept as others'. However, the morale of the people had improved 'though to have said that the spirit of insubordination was altogether overcome, that the people were as calm and peaceable as their English fellow-tenants on Mr Spencer's estate in Berkshire would be absurd'. The two nations were so different that that could never be, but the Irish tenants had 'learned that their landlord really cared for them – that he wished their worldly, as well as their moral, and mental advancement, and they were not dissatisfied – they had a stake in good order'.[23] This, unfortunately, is fiction and an expression of what Mrs Hall wished to see. In Ireland, in reality, the vision of the contented tenantry prospering under the benevolent eye of an equally prosperous landlord, learning from him, and profiting by his good example, was as far from being the norm in real life as it had ever been.

Mrs Hall touched on an interesting subject in her novel – that of the old Irish owners of the land. In *The Whiteboy* it is the Macarthys who are the last remnant of the Gaelic aristocracy and Mrs Hall paints a rather affecting picture of a great lady who remains true to her family traditions and dies in the ruined remains of her family home. Her grandson, Lawrence, is doomed, as there is no place for him in the new world of the English landlords, and he cannot forget his heritage. He is seen: 'his hot high blood beating in every vein; with all the pride and all the achings after Nation, that belong of right to old descent; yet shut out by the then existing laws from nearly every path that led to honourable distinction'.[24] The peasantry, too, are shown as resenting the Saxon usurper who lives in the lands that are rightly those of the Gaelic chieftains and as loving and respecting Lawrence because of his lineage.

In all her previous writings Mrs Hall had shown an admirable tolerance of religious differences, while making it clear that she was of the 'Reformed faith' she had expressed an admiration for certain aspects of Roman Catholic behaviour. Unlike the rabidly anti-Catholic writers, George Brittaine or Charlotte Elizabeth, she did not portray priests as sinister figures and had spoken up on their behalf in several of her writings. Her earlier commendations were tepid, however, in comparison to the warmth of this passage in *The Whiteboy* analysing the relationship between priest and parishioner:

> There are few things in the actual world so touchingly beautiful as the respect and affection subsisting between the Roman Catholic priest and his flock; those who know and observe the people cannot wonder at their strength and endurance; from the cradle to the grave the priest is the peasant's adviser and his friend; he knows all his concerns – not only the great business of his life, but its *minutiae*; his private cares and sorrows, his faults and his crimes, are all in the priest's keeping; his judge, his advocate, his punisher, he is also his protector – very, very rarely his tyrant.[25]

The emphasis placed on 'crimes' signified that Mrs Hall wished her readers to know that she was aware of the fact that there might well be collusion between priests and people and that her toleration did not go so far as to condone treason. Yet she still struggled to explain the causes of treason and disaffection instead of simply condemning their effects. Dean Graves, the local parson, explained his attitude towards Catholics to the new English landlord, Edward Spencer, within a short time of meeting him. Edward says of Catholics: "You cannot call their faith insulted now", and the Dean replies, "Were my faith so treated in this year, 1822, I confess to you, I should so consider it". He indeed, wished that all Irishmen could be Protestant, but he deplored the 'wickedness and folly' of the penal statutes which had constrained the Irish Catholic church. His views as a minister and as a humanitarian were in conflict, and Mrs Hall's own views are expressed by him when he says

> "Though, as a clergyman I would joy to receive those who felt themselves in error, I cannot, as a man, but respect the firmness which, despite scorn and degradation, has hitherto united the peasant to his faith, mingled though it be with a superstition which is part of the national religion, as well as the national poetry."[26]

Dean Graves was Mrs Hall's ideal clergyman, literate, articulate, tolerant and thoughtful. Not all of his qualities were appreciated by some of the other characters in *The Whiteboy*, and in the gathering at Macroom Castle, the country gentleman, O'Driscoll, spoke slightingly of the Dean's tolerance: '"Only I have such a respect for the Establishment, I'd never enter his church; He's not worth a farthing; he's no spark in him; he's a *moderate man* – think of that! Sure, I once heard him call a priest his Christian brother'".[27]

The Catholic priest in Dean Graves's district is very sketchily drawn; it is merely established that he is good, holy and kindly and in no way does he form an artistic balance to Dean Graves. There is, however, a Catholic priest called Father Jasper, who makes a few appearances and who is presented as scholarly, with a particular interest in Irish history and antiquities. He is a refined man, who had been 'educated at St Omer, and was fonder of introducing scraps of French into his conversation than shreds and patches of Latin'.[28] In complete contrast is the wicked agent, Abel Richards, the villain in *The Whiteboy*. He had been converted from Catholicism to Presbyterianism while a boy, by Mrs Spencer, the aunt of the hero, Edward Spencer, when she had lived on the estate in Cork, and had prospered in his new religiion. By clever scheming he had acquired property and was a hated landlord and middle-man. He is shown as both cruel and cowardly; he trembles and grovels in fear when he hides from the wrath of the peasants, but then betrays the heroine, Ellen, who had saved him. In no sense is he a believable character, for he possesses not one redeeming feature, and his part in the action of the story is peripheral. He could be removed and it would make little difference to the plot. Yet, without him, the book would be much poorer because he is the vehicle for one of Mrs Hall's didactic aims – the promotion of religious tolerance. Abel Richards is a caricature of the proselytisers who exacerbated tensions in Ireland at the time, people who did a great deal of harm and achieved no perceptible good, and of the proselytised who allowed self-interest to take the part of conviction. Abel's 'sanctity' is ridiculed as, without any self-mockery, he speaks of himself as "a poor sinner – a humble, faithful servant of the Lord". He had 'holy words' and 'polluted lips', and moved about 'forever seeking to distribute the tracts which he kept in his pocket. 'Low-hearted creature as he was', he was

> possessed of considerable dramatic talent ... His hands were folded, his head was bent, his eyes were half-closed; his was the martyr aspect of one who ... had submitted to the scourge and the cord, and sundry beatings of the breast, and resolved to carry his stripes into society and descant upon them as people do upon good fortune.[29]

Richard's speech was a 'low tone of vulgar cant' and included such phrases as '"highly-favoured"', '"brand from the burning"', '"humble martyr"', '"Christian forbearance"', and '"butcherly Papists"'.

Mrs Hall has much amusement from her creation, and obviously enjoys her mockery of those who practise sectarian religion for their own purposes, but she fails to establish Abel as a menacing figure in whose 'impure' mind 'a multitude of evil thoughts were in perpetual motion, like tadpoles in a stagnant pond'. The simile is unfortunate – it conveys no idea of horror or terror, but is faintly comic. Much more rewarding is Mrs Hall's description of how certain sections of society reacted to Abel Richards and his cause:

In Dublin he dined with titled ladies, learned to eat with a silver fork, obtained presents of bitterly-worded tracts from those who had the reputation of sanctity in their own "set" – other, more timid votaries gave him blue and pink book markers embroidered with words which ... were at decided variance with their practice, – thus a lady who would not suffer a "Popish" domestic to enter her service, selected the motto "Charity suffereth long and is kind", and another, the simple word "Peace" worked in *orange* silk as a token of her hatred of the green.[30]

The rural gentry who 'scorned and avoided' him were more culpable, because by their tolerance of him they sympathised with sectarianism. Nobody disowned him, or publicly disapproved of his actions, because, whatever his faults, he was one of "their party", and solidarity in the face of common danger was all. As events move on, however, in the story, the gentry distance themselves from Abel Richards for he makes the mistake of slandering the heroine, Ellen, whose feminity and family connections awaken both the chivalrous and the class instincts of the gentlemen. In the end Abel Richards dies a violent death, and his body is reclaimed by his mother, who mourns over it in the traditional Irish peasant way, so that a neat conclusion is reached and the intruder who had been foolishly allowed into Protestant Ascendancy society is returned to his rightful place. Consciously, or unconsciously, Mrs Hall is telling us where she stands on the problem of class, always a preoccupation of hers.

Religion, in Mrs Hall's writing, is not a question of dogma, rather it is, as Barbara Hayley has said of Gerald Griffin, 'a code of practice, a guide in daily life'.[31] Master Mat, the schoolmaster in *The Whiteboy*, expresses it well when he speaks to Ellen of the Protestant religion they both practise: '"Your religion, dear, is in hourly activity – bearing and forbearing – doing deeds of love and charity towards your fellow-creatures, because they *are* your fellow-creatures, and God's creatures as well – that's it"'.[32] This all-purpose belief would serve equally well for Protestant or Catholic and Mrs Hall seemed to indicate that there was no virtue in exchanging one formal religion for another.

The Whiteboy is the story of an Englishman's awakening to the realities of Irish life and Edward Spencer suffers many shocks on his journey to comprehension. Irish poverty, a recurrent theme in all Mrs Hall's writings, is beyond anything he could ever have imagined. When he arrives in Cork he is surrounded by beggars whose aspect horrify him – his first glimpse of Irish destitution:

In England, residing in an agricultural district, he had only seen poverty in what might be called its "pictorial form" – in the person of an aged widow receiving her weekly alms – or under the aspect of a group of rosy children, who, but that you were told, so would not have been imagined orphans ... Alas! Alas! those who ran wolf-like, yet harmless, screaming after his car, seldom tasted animal food once a year. All they complained of was, that the old potato-crop was exhausted, and the new not yet come; and till it did, they must starve, "only for charity." In England, poverty, at that time more particularly, was so great a shame that it hid its face and wept; but in Ireland, it breasted the breeze and howled in the sunshine.[33]

In England vagabonds were committed as rogues under the vagrancy acts, and poachers were punished for their crime, but in all cases their 'poverty was alleviated'. Mrs Hall has now moved from blaming people for their poverty to blaming the authorities for not dealing with the effects of it, but there is no question of examining the root causes of that poverty. At the conclusion of the book there is an idyllic scene of peasant contentment and prosperity, achieved by following the wise precepts and good advice of the new young landlord, and the only problem left to solve is that of stubborn tenants who still cling to their old lazy ways.

The Irish peasant spent most of his life poised on the edge of starvation and in some districts actual famine was endemic. There were two particularly bad famines in the early years of the nineteenth century – one in 1817, the other in 1822. One wonders why 1822 should have been chosen for the year in which The Whiteboy is set, because although Mrs Hall's plot depends on the twin dangers of Whiteboyism and French armed intervention, 1822 was not marked by increased agrarian crime or by foreign invasion. It is then a melancholy coincidence that The Whiteboy should have been set in a famine year, 1822, and published in the first year of the greatest famine the country had yet known. The novel appeared in August 1845 and the first effects of the potato blight were not noticed in Ireland until October of that year so that Mrs Hall's choice of a famine year *was* nothing more than a coincidence, but her descriptions of the earlier hunger had a topicality that she might not have anticipated. These descriptive passages are extremely moving, marked as they are by a restraint that is more powerful than impassioned outbursts of authorial feeling. The scenes of dearth and disease are seen through the eyes of Ellen, the heroine of the novel. She had been ill for months and had not known what was happening in the countryside all around her. She was familiar with poverty, hunger and disease among the peasantry although her position in Spencer Court protected her from personal affliction, but this was far worse than she had ever seen and 'a cold shivering crept through her frame as she observed the cabins one by one that were scattered through the glen, and saw how few signs of life were moving about them; the tell-tale smoke did not arise from more than two or three'.[34] A funeral was taking place, attended by 'poor spectral creatures escaped for a little while from the starvation which, if it continued much longer, must overtake themselves'. Elsewhere there were women keening over the dead, and unfortunates stricken with fever were lying in ditches under makeshift shelters.

Most poignant was the state of the deserted schoolhouse that showed all the signs of months of neglect. Weeds were growing through the flagstones of the path, the woodbine was hanging down untended, and a robin had built its nest in the broken waterspout. The schoolroom itself was damp:

A large yellow frog had been squatting in the hearthstone, and with its stony eyes, and bloated sides, slouched heavily away as Ellen entered. Crickets were lying dead among turf ashes. The cupboard, were the boys ought to have kept their slates, but where Master Mat stored away his "eatables" in former days and which had been full of little fat, frisking mice whom he could never bear to destroy, was open, and silent. The blackbird ... that whistled his wandering master up many a morning at daybreak, had died of starvation; his yellow bill hanging over the broken blue tea-cup that once contained its food. The long tables looked like spectres of fir and deal. The books ... were damp and mildewed; some piled up, others were scattered on the floor ... The spiders ran along their huge webs, or crouched in the centre of the filmy architecture, astonished and terrified at the vibration of a human voice.[35]

This description of an empty, desolate spot is more powerful an evocation of death and misery than a blunt cataloguing of human suffering. The small, trivial details accumulate, and the inappropriate gaiety of the blackbird's golden beak and the bright blue of the tea-cup jar against the sombre surroundings. Nature has taken over, the weeds proliferate, while everything else decays. It is a metaphor of the state of Ireland – once happy, busy, full of life and gaiety, now silent, still, and slowly dying. Is the blackbird a symbol? It could have fended for itself if it were free, but caged, it needed to be fed and when there was no hand to feed it the bird died.

A starving and subject people will, naturally, blame their masters for their plight, and in Ireland an added edge of bitterness was given to the blame by the fact that corn and cattle were being exported from Ireland while the poorest of its people were starving. Mrs Hall had already highlighted this public scandal of the hungry years. In her *Stories of the Irish Peasantry* (1840) Mrs Hall laments that in the desolate and lonely city of Dublin the only sign of activity is on the quays: 'corn and cattle may be seen there in abundance, but both are on their way to England; they pay no duty ... The corn and cattle are to be exchanged in British markets for money which the Irish farmer is not to see; it passes from the hands of the 'driver' into those of the banker, to my Lord This and my Lord That'.[36] In 1845 in *The Whiteboy*, Mrs Hall again referred to the trade when her English hero, Edward Spencer, in his brief pause upon the quays in Cork 'marvelled that a hungry population allowed droves of cattle; crates filled with eggs and poultry – all the evidences, in short, of surplus plenty, to leave the country without hindrance'.[37] There were, in fact, food riots (referred to by William Carleton in his famine novel *The Black Prophet*, published in 1847) but none that were not easily contained by the authorities. The sad spectacle of ships laden with Irish provisions leaving a country full of starving people was a result not so much of English governmental malice or indifference but because of the official belief in Free Trade which regarded as sacred the workings of the private market. Some modern economic historians have asserted that the export of Irish provisions made little or no difference to the outcome of the famine in Ireland, but that, of course, is not how it looked to those who were starving, and Mrs Hall was expressing the bewilderment and resentment of those who knew nothing of economic theory, but far too much about what happened in practice.

The Whiteboy is a novel about land and its ownership, and as Mrs Hall knew very well, Irish crimes of violence were almost always connected with land disputes or grievances. The book is a serious attempt on her part to come to terms with those Irish problems which had been caused by violence, whether legitimised or not. As an author she was no stranger to violence – a strong streak of it runs through all her works, and with it she caters for contemporary tastes – but as a moralist she deplores it. By setting *The Whiteboy* in 1822 she was free to comment on Irish affairs as if they were a matter of history and not of contemporary fact. The choice of title was a good one, because the term 'Whiteboy' was in popular use to denote all types of agrarian violence in Ireland. It would have been generally understood what the subject of Mrs Hall's novel was, in a way which might not have been possible had she entitled it *The Caravat, The Shanavest* or even *The Rockite*. Also to be taken into account is the fact that George Cornewall Lewis in his influential book on agrarian violence, *On Local Disturbances in Ireland*, (1836), had argued that the Whiteboy movement had come about in response to an exploitative land system and that the main sufferers were the peasants themselves. This theory would have accorded well with Mrs Hall's belief in the self-defeating effects of violence and also with her conviction that the Irish people were incapable of seeing where their best interests lay.

There is a great deal about violence in *The Whiteboy*; not only the violence inseparable from an armed rising, but the violence sanctified by law and government. A servant girl says bitterly to her mistress, Ellen, who has exhorted her to respect the law: "The poor never hear of law, except from the man that distrains their bit of land, sends them to gaol, or houseless and homeless through the world!"[38] and the author herself comments acidly on government policies in Ireland and methods of dealing with unrest. Speaking of the 'disturbances' of 1822 she says:

At last the higher members of the community, the magistrates and "the castle" conceived it necessary to do "something", and thus originated the meetings at Macroom, Kenmare, and various towns throughout the country; and the "something" done in 1822, was just the "something" done during the whole period of English rule over Ireland previous to that year; namely, placing the country under martial law, burning and hanging, and shooting those whom fever and starvation had spared. And doubtless, when the hunt was closed, and the "dogs of war" were whistled back; when the judges were wearied with passing sentence, and the ships heavy with convicts; when there were many widows and orphans in the disturbed districts – then "the troubles", as in former days, would be pompously announced as "over"; the country reported as "tranquillised"; no effort made to employ and pay those who were growing up to manhood with no other occupation than the memory of their sire's wrongs; no hope beyond hunger, revolt and death.[39]

When Edward Spencer, the liberal and fair-minded English heir to an Irish estate, is travelling towards his property he comes upon evidence of the 'disturbances' and their repression. He sees the 'blackened and smouldering' walls of a cabin which had been burnt by the military, and further on 'a still more distressing evidence of martial law; a group of women were wringing their hands, howling and weeping most bitterly over the inanimate form of a man who had evidently met with a violent death'. According to the witnesses he had been roused from his bed by the soldiers and shot without trial in front of his wife, '"whose arms they untwisted from round his neck – *with a bayonet*"'. Edward saw for himself that this was true: 'for though she was in a swoon on his body, her hands were cut and bleeding'.[40] This was no act of justice in Edward's eyes, but an act of violence, and he spoke eloquently about it when he met the Grand Jury at Macroom Castle, but was shouted down.

There are a couple of references to casual violence being inflicted on the lower classes by their superiors; old Mr Spencer, Edward's uncle and owner of Spencer Court had been heard to grumble that 'the representative of a certain ancient family had actually committed a breach of the peace by knocking down his own coachman', and during the tumult that followed the arrival of two funerals simultaneously at the churchyard (one would not give way to the other because of the belief that the second person to be buried must tend the first one in Purgatory) the Protestant party dealt roughly with the Catholics. In Mrs Hall's words: 'Many of the gentlemen rode in among the people, beating them down with their heavy riding-whips and trampling them beneath their horses' feet ... and it was evident that there were many ... who would not endure the blows – which the rich in those days were sufficiently apt to inflict upon the poor'.[41] Although actual descriptions of gentlemen flogging peasants are not a feature of Mrs Hall's work, the potential for such violence is there. Peasants are shown as humble, ingratiating in speech and servile in behaviour, while their superiors speak loudly and use arrogant language. Mrs Hall, however, attributes the greater violence to the Whiteboys, and Murtogh, the 'half-savage', and his foster-brother Lawrence are shown as the types most representative of the members of the organization. Murtogh is a peasant, and Lawrence is a descendant of a noble family but they are united by their common love of the land and hatred of those who stole it from them. Murtogh is a crude version of Lawrence, and is more forthright and direct in his search for vengeance, but basically they are the same because they live by violence. Lawrence commands a band of outlaws and leads them in ambushes, but does not himself take part in burnings, maimings and assassinations. Yet he condones them, and so is as morally culpable as Murtogh, who has himself taken a leading role in the crimes of violence. Murtogh, indeed, tells with relish of how he killed a land-grabber, not without difficulty, because the victim's children appeared without warning to greet their father as he came riding up the road: '"Tree or four of de young ones had got in de grove o' trees and put dere little faces over de ditch shouting 'Fader, fader'. And he pulled up and tossed over his hat among 'em, and it was half-full of playtings for dem; and faix I'll not deny dat de shout of dere joy made meeself uneasy, just at the minit"', but Murtogh shot the man nonetheless. He did not see this as murder, but as justice, and was horrified by the suggestion that he might have shot the youngsters: '"Shoot de childre!" he exclaimed,"... *is it murderers ye'd have us be?"*.'[42]

Murtogh dies a violent death, dragging down into the sea with him the wicked middleman, Abel Richards, who had been himself guilty of many crimes against poor people, and Lawrence too dies but there is a difference in the manner of their death. Murtogh's was sudden and the result of a conscious and impetuous decision of his own, while Lawrence died of a wound received in a battle with the forces of law and order, having had an opportunity to be reconciled with his half-sister Ellen. It was a death more in keeping with a gentleman, even a half-gentleman, than Murtogh's would have been. Both men lived by violence and both died by it, but a third, the exiled Irishman, Louis, who had come from France to work with the rebels and to prepare the way for French military help, was allowed by the author to escape, although seriously wounded. There was justice in his fate too, because he had not approved of the Whiteboy methods, and had spoken out against them saying

"Every one who stabs in the dark, or fires from his concealment upon the unwary, IS AN ASSASSIN. No matter what his cause may be, he is the thing I have named! I did my best to prepare you for a bold struggle, man to man; to the last gasp of my existence, I would fight for that; I would aid a guerilla warfare – it is the refuge of a mountain people for their defence; but the stab in the dark; the man shot into the very arms of his children, hundreds of miles away from the scene of action, for disobeying laws as arbitrary as those you complain of, to which he never gave his sanction, and which he was never bound to obey, is a thing that cries for vengeance and brings a curse instead of a blessing on our cause!"[43]

In this speech Louis is voicing the views implied by Mrs Hall throughout the novel, especially the conviction that secret societies with their oaths and rules and the punishments for infringement of the rules, are harming the cause of the peasants (the Cornewall Lewis view). She illustrates the effect upon the innocent individual by having the schoolmaster, always a conservative and kindly figure in Mrs Hall's works, unlike Carleton's schoolmasters, tell of the murder of a post boy. The lad had boasted that he was not afraid of the outlaws, and they, hearing this, shot him as he went on his rounds. Master Mat describes how the mother went for the body:

"She drove the car herself, poor thing; the gray rocks hanging over her road, and the fox and the eagle looking down upon her trouble; and she found him with a blanket thrown over him, but nothing living near him; the rifled bags and torn letters, and his dead horse were there too – though she didn't heed them; and God gave her strength while the stars were shining above her, to lift the body into the car – and there she sat, alone with her dead!"[44]

It is, however, the overall effect of the Whiteboy movement on the Irish people that most concerns Mrs Hall and she is eloquent in her descriptions of the reactions it produced. She, or rather, her principal character, Edward Spencer, notes how a mob, transformed from the 'helpless people who that morning had whined forth their solicitations for '"a halfpenny, for God's sake"' had been checked in their harassment of a party of soldiery by 'flushed, heated and strange men ... who grasped their shillelas with strength and determination'. These, he was told, were Whiteboys who were organizing a diversion, and their success was proof of their power over the people. This power was achieved not only by the rituals of oath-taking and the threats of punishment, but by the natural leanings of Irish people towards violence. They came to Whiteboy meetings

instinct with life, and the great purpose of Irish life – action. Something was to be dared and done, and that at once; this was enough for them; enough at all times. The ready and willing instruments of danger and defiance, with heads throbbing as wildly as their hearts, every nerve thrilling with emotion – they came, they gathered around their leaders, all doubt and coldness vanished, eager and panting for action, for movement, for anything rather than reflection and repose.[45]

The denigration of Irish motives for rebellion that was noticeable in Mrs Hall's earlier works is now fully established, and her view of the Irish as careless, reckless, and, most of all, unthinking people is clearly visible. The peasants in her early sketches who were too heedless to mend their roofs, or pen up their pigs, or fence their gardens are now more dangerous because they have fallen under the sway of agitators who know how to tap their unthinking violence. They had 'no defined idea of a struggle for their country – it was but a piecemeal effort to keep their bit of land', and were 'incapable, as unwilling, to reason'.

Although Mrs Hall is aware of the dangers of the Irish propensity to violence and of the eagerness with which unthinking men will listen to demagogues, she remains optimistic of the great good which can be done by caring landlords who live among the people and who understand their natures. The Master of Macroom is such a man, rough, bluff and hard-drinking, a good Orangeman and a bigot, yet he knows how to influence a mob. He addresses a hostile gathering of his tenants saying:

> "Nothing shall change me but your own violence and disregard of the laws of your country. I warn you most earnestly not only against any act of violence, but against assembling in a manner that will lead to violence. If you commit an outrage I cannot save you. Go home peaceably, and by so doing prove that no outrage can be laid at your door; go home if you would save yourselves and your families from ruin."[46]

Mrs Hall emphasises here yet again that the people are themselves the greatest losers by violence, and that those against whom it is directed are, in fact, their natural protectors and true leaders. The main thesis of *The Whiteboy* is that the country will prosper if its affairs are run by resident landlords responsible to a political assembly in London, and this prosperity will bring peace and contentment to all. Violence threatens progress towards this aim – violence by the authorities which causes anger and resentment among the peasants, and the violence of the peasants themselves against the landlords. Her argument is that the efforts of the local, far-seeing and kindly landowners to run their estates properly and to see to the welfare of their tenants are being nullified by the brutal and repressive measures of the Government, but she has no answer to the question of how the attitude of the Government is to be changed. Peasant violence can be limited and contained, and local grievances can be remedied by the efforts of benevolent landlords, but who can check the violence of the law?

From the very start of her literary career Mrs Hall had made it clear that she wished to be thought of as completely non-political and in one of the tales, 'Too Early Wed', which appeared in the collection *Stories of the Irish Peasantry* in 1840, she stated firmly 'You will find no politics in my sketches'.[47] It is difficult, however to discern what she means by 'politics' because on the great political question of the early nineteenth century in Ireland, the repeal of the Union, she takes a clearly-defined position. She is firmly anti-Repeal, and pro-Union because of the benefits it will bring, in time, to her native Ireland. Her stated aim in writing her Irish sketches was to make Irish people known and loved in England and to correct the native failings which would prevent the Irish from enjoying the benefits of Union. It would have been strange if Mrs Hall had managed to write about Ireland without touching upon politics and if she had done so she would have been the exception among Irish writers of the day. Her contemporary, Lady Morgan, set the tone when only five years after the Union, her novel *The Wild Irish Girl* (1806) celebrated the legislation in a symbolic marriage between the English hero and the Irish heroine. The book was pro-Union, and the author's enthusiasm for the measure was expressed in the words of the hero's father who saw it as '"being prophetically typical of a national unity of interests and affections between those who may be factitiously severed, but who are naturally allied"'.[48]

A completely antithetical point of view was put forward in Charles Maturin's novel *The Wild Irish Boy*, published in 1808. The hero, Bethel Ormsby, who had found in Ireland the romanticism which had eluded him in the Lake District, saw the Union as a disaster for Ireland. It encouraged absenteeism among the nobility and gentry, it had 'diminished her population; impoverished her metropolis [and] depressed her internal business'.[49] These two opposing views were, in various degrees of conviction, shared by Irish writers for years to come, giving force to the comment by the critic in the *Athenaeum*, reviewing the O'Hara brothers' novel *The Anglo-Irish of the Nineteenth Century*, that 'Irish novelists have uniformly given a party cast to their fiction ... It is certain that not one able Irish romance, from *The Absentee* to *The Croppy* has been unmarked by this peculiarity'.[50] The use of the adjective 'able' is of interest: it implies that political matters are by no means unsuitable material for a novel, and may indeed add to its interest, and 'peculiarity' stamps the works as being outside the normal scope of English novels. 'Political' is interchangeable with 'national' in this context, because whereas English politics were an internal matter, Irish politics had to look to the outside, to the colonising power.

What distinguished Mrs Hall's political didacticism from the lessons taught by other Irish writers of the time is that she is preaching the essential inferiority of the Irish people. Lady Morgan speaks of a marriage, which implies some equality of position, but there is no hint in any of Mrs Hall's writings of parity between the two nations: the Union is not one of equals but of superior and inferior; indeed it is not a union so much as an absorption by the stronger of the weaker. Her concentration on the faults of the Irish people, their indolence and fecklessness, is an illustration of her belief that the Union must stand, and even the virtues she grants to them, generosity and kindliness, are not those which are highly rated in a political sense. Irish peasants' other great virtue, fidelity to their masters, proves that the Irish are happier when they have someone to lead them, than when they have to find their way on their own. Again and again in her sketches Mrs Hall stresses this point, and sets up good resident landlords as the natural leaders and protectors of the people. Paddy does not *think* but *feels*, Mrs Hall believes, and he acts upon instinct, and a good landlord knows how to guide him. By extension then, an English government which exercises patience in the management of the difficult Irish nation will succeed in winning the hearts of the people.

Unfortunately, not all English governments had shown this patience or had behaved towards Ireland in a generous and tolerant manner, and in *The Whiteboy*, Mrs Hall rebuked the English politicians who had failed to implement the political union in a way which could have benefited Ireland. It is in this book that she clearly enunciates the political views she had once merely hinted at or had conveyed in a more oblique way through morality stories teaching how English virtues could transform miserable Irish lives, and in it she tackles the whole problem of Irish nationalism. Her hero, Edward Spencer, is educated by the people he meets, and by his experiences in different sections of Irish society. In him Mrs Hall creates the ideal Englishman; tolerant, fair-minded, willing to suspend judgment on local behaviour, and ready to learn as much as he can in order to understand the motivations of the people among whom he has come to live. He is also wise, philanthropic, brave and chivalrous, and above all, firm, steady and well-balanced. In contrast, the Whiteboy of the title, Lawrence Macarthy, the poor and landless descendant of a once-great Irish family, is given to wild talk and even wilder action, allowing himself, because of his grievances, to be led into joining the illegal organisation which is terrorising the district. He is brave but rash, and is lacking in honour because he involves his half-sister, Ellen, in his troubles, and endangers both her physical safety and her good name. She is rescued in both senses, by Spencer, who then marries her and brings her to a life of material and emotional comfort.

In assessing the lesson taught, the reader should realise that Ellen is not native Irish, but of settler stock and of the Protestant faith, brought up in the home of a landowner and reared to respect English values. (She and Lawrence were half-brother and sister, Lawrence's father having died while the boy was very young. His mother had married again, this time to a Protestant.) Lawrence, on the other hand, has been brought up by his native Irish grandmother, and is a Catholic. He is killed in an abortive uprising, a demonstration of the futility of resisting English rule, which must inevitably prevail, because of its superior virtue. Dean Graves, the Protestant clergyman who enlightens Edward Spencer about many aspects of Irish life, speaks sympathetically of the past injustices suffered by Irishmen at English hands, especially the confiscation of their lands, and points out that the Irish peasant is ever mindful of his lost legacy: '"wrenched from him by force or fraud"'. In his opinion, however, the problems of Ireland would not have become so difficult to solve if there had been material advancement for the Irish people; '"if they had been drawn into active life; if they had found their labour sufficiently productive to afford subsistence; if efforts had been made to elevate, and not depress them, in the scale of humankind ... the poor Englishman clings to his comforts; the poor Irishman might have done the same, *if he had had them to cling to*"'.[51] Edward takes note of this advice, and the story concludes with a vision of his Irish estate transformed by his good works, where happy peasants live in comfort and contentment, no longer cold, hungry or idle under the new dispensation. There are still a few tenants who have not yet understood the value of the improvements but Edward and Ellen are making it their lives' work to bring the benefits of material success to all in their care.

The Spencer method in dealing with dissidents is recommended to his countrymen by Edward when he beseeches his wife: '"Let us endeavour to persuade England to try – or rather, to continue another course; by showing the POLICY OF GENEROSITY, and THE WISDOM OF JUSTICE; and so RECONCILE THE DOUBTFUL, OR THE SUSPICIOUS, TO THE MORE LIBERAL SYSTEM WHICH MUST BE ADOPTED IN GOVERNING IRELAND HEREAFTER"'.[52] Ireland must be 'governed', the Union is a permanent reality. Mrs Hall is using Edward to express her views on Repeal, reiterating yet again the importance of the material blessings the Union confers, and envisioning a bright future if people will only work together for the good of the country:

"There are men who unfortunately consider that now emancipation is obtained, they must spend the other half of their lives on a project which they themselves cannot with all their enthusiasm, consider feasible. This, according to my views, is deeply to be regretted, because the same energy and perseverance directed to the real improvements of the country would create and multiply practical blessings".[53]

Such men, supporters of Repeal, who cannot see that England is no longer their enemy, but their friend and guide, constitute the real danger to Ireland's peaceful development. The country "has now far

The Whiteboy – 'A truly national novel' 171

more to dread from internal disease than from external oppression", and because of the force of public opinion in England, and its insistence that justice be done, '"England dare oppress Ireland no longer – not from fear of Ireland, but from fear of England"'. The possibility of an alliance between the two chief parties in Irish life, Catholics and Protestants, is no longer given any credence, in spite of the Protestant clergyman's earlier warning to Edward Spencer of such an alliance when he spoke of '"Protestant Ireland and Catholic Ireland, Saxon Ireland and Celtic Ireland"', where '"the blood of the Irish Saxon is as hotly boiling as that of the Irish Celt. They boil against each other; and perhaps the chief motive of union between the Saxon Irish and your country is the Church establishment; that is the bond which binds the Irish Protestant to England, but for that both might overboil against you as they now do against each other"'.[54] The gulf between the parties in Ireland, however, is so great that this will never happen, and party feeling is manifested in even the most trivial incidents of daily life. Edward Spencer had been warned of this, when he first arrived in Ireland and told his shipboard acquaintance, the Catholic Lady Mary O'Brien, of his intention to live in Ireland without party affiliations. Lady Mary advised him that he would

"not be permitted to do as he pleases; to dream of being allowed to steer a middle course undulating between orange and green, with a leaf of our southern shamrock in one button-hole, and a petal of the northern lily in the other, dancing down the middle with an Orangeman on his right hand and a Whiteboy on his left, then up again with the Whiteboy on his right, and the Orangemen on his left. It is really too ridiculous".[55]

In spite of the warning, Edward made what were regarded as mistakes. His bottle-green cloak and the grass-green livery for his servants not only caused confusion, but gave offence to one section of the population. He had chosen the colour as a compliment to Ireland and 'a delicate way of evincing the interest he felt in the Irish', but, as he sat in his Protestant-owned hotel in Cork, having ordered the Catholic newspaper, to the consternation of the staff, he 'felt already that he had made a mistake, that green was only the natural colour of Ireland, *out* of Ireland, and in Ireland the colour of one party; the substantial shebboleth [sic], the palpable war cry as it were'.[56] In *The Whiteboy*, Mrs Hall is more concerned to emphasise the material damage done by politics, because they threaten the Union, than she is with their danger to personal morality, but she does convey a measure of contempt for the importance attached to trivial gestures of partisanship. Thus, Edward Spencer, because of the nobility of his nature, is described as being 'peculiarly unfitted to comprehend the bickerings and bitternesses, the petty malignant nothings which from the superstructure of Irish discontent, frequently to the exclusion of thought for actual and positive grievances'.[57] The plot of the novel is unbalanced because nobody can do anything properly except the English hero. The central Irish character fails in his rebellion and is killed, his aristocratic companion in arms, another Irishman, has to flee the country, and his native Irish foster-brother is drowned while trying to escape the pursuing officers of the law. The characterisation of the peasant is crude and blatant caricature. He speaks in broken English, and is described by the author as being 'few removes, perhaps, from a savage; keen, cunning, revengeful, cruel but faithful, and undeviating in his few attachments – superstitious to positive weakness; his dark, deep, blood-shot eye, that never quailed at difficulty, would tremble amid tears of terror at the mention of a ghost, or a tale of supernatural agency'.[58] Such a creation could not be expected to be capable of anything more than animal responses, but the Irish Protestant gentlemen who sat on the grand jury in Macroom might be expected to show intelligence and initiative.

This, however, is not the case and they are at a loss to know how to deal with a potential rebellion until the Englishman, Edward Spencer, shows them the way. At a meeting to discuss what action should be taken, where 'Irish indignation, and the rash, impetuous nature of Irish eloquence, with its mighty torrent of invective – of appeal – of destruction – of revenge, were poured forth' the young Englishman sat silently and his manner gave the impression that he did not agree with the repressive methods proposed, but eventually he spoke, and such was his cool commonsense and practicality, that the company 'deferred to him as if he were an old resident' and he helped lay the plans that outwitted the rebels. Such was his evident superiority, not to mention his manly courage, that the very Whiteboys were impressed by him when he insinuated himself into one of their secret gatherings, and he was not harmed in any way, but was kept prisoner and well looked after by them. This neardeification of Edward, the Englishman, is a major flaw in the novel. There is little dramatic tension in his conflicts with others, for he is always destined to win. His character, that of the good man, is as overdrawn as that of Abel Richards, the wicked man, and he must be seen as no more than the personification of what Mrs Hall considered the ideal landlord to be.

The whole tone of *The Whiteboy* is buoyant and hopeful. While she was writing it Mrs Hall perceived Ireland as a country where great changes for the better had taken place – the Tithe War, which had brought real want, hardship, danger and even death to ministers of the Protestant faith and their families, was over and the religious problem seemed at last to be soluble; Father Mathew's temperance crusade was a resounding success, agrarian unrest was quelled, and the arch-enemy of the Union, Daniel O'Connell, had lost some of his popular support when his much-vaunted Repeal Year of 1843 failed to produce the abolition of the Union. Solutions to all Irish problems seemed to be in sight, if only the Irish people – and the English government – would listen to what Mrs Hall was telling them. Reviewers in Ireland and in England took the novel and its message seriously. English reviewers perceived *The Whiteboy* as a political novel, and the Catholic periodical the *Tablet* went so far as to link Mrs Hall's name with that of the Repeal party, thereby missing her point, that the Union was the only hope for Ireland. 'We are not all inclined to discuss a political question with a lady', stated the reviewer, yet, 'there is much in her book to amuse and instruct, and as its purpose is excellent, however we may differ as to the means by which she proposes to work out the great end she has in view, we hail her *Whiteboy* with a hearty welcome. "Agitation" is one of the horrors of the amiable writer, and yet her novel is a speech in Conciliation. Hall expanded into two volumes; the facts and the substance are common to herself and the Repealers and they have the same object in view.'[59] The literary critic of the *Spectator*, describing the novel as 'didactic', praised Mrs Hall for having 'thrown a new light upon the cause of the "national" feelings that irritate the mass of the people and the Young Ireland party'. It is of interest to learn of the problems of Ireland, and the reasons for them, but he concludes that the remedies Mrs Hall proposes are 'not very practicable; for they seem to consist in a resident landlord, backed by an English fortune, improving his estate and his tenantry'.[60] It is a measure of how seriously Mrs Hall was taken in her time as a writer on Irish issues that this review was quoted in O'Connell's newspaper the *Pilot*, in August 1845, in an article on the question of confiscated Gaelic estates. Referring to her views, expressed in the *persona* of Dean Graves, on the subject of land ownership, the writer opined that she knew little about the matter; that such ownership was no longer in dispute, and that any small practical problems that arose from land transference could be easily solved.[61]

According to *The Sunday Times* the novel was Mrs Hall's best story to date, and it noted that the work had a high purpose: 'intending to expose the defects of the moral and political condition of the sister island, and to show how banefully the mistaken notion of regenerating that country by the Bible and the bayonet has worked'.[62] Here was an acknowledgment of Mrs Hall's aims, and an implicit acceptance of her view of the Union – peaceful progress through prosperity instead of repression and coercion. The reviewer in the *Literary Gazette* praised the description of the evils which afflicted Ireland: 'that excitable, uncertain, mercurial and hard-to-govern country',[63] and commended Mrs Hall for her clear analysis of the problems, but did not care to say whether or not she had enabled readers to solve them; they would have to decide that for themselves.

The Whiteboy was greeted by the *Athenaeum* as 'indisputably her best' 'with eloquent description' and simplicity of plot. Her talents as a novelist, however, are secondary to her high purpose:

She is, after all, but a national teacher; since while her lips are ever busy to inculcate the common-sense and clear view of practical duty which Miss Edgeworth was the first to hold up to the Lantys and Rackrents of the dear country – her heart inclines towards that romantic devotion of high spirit to picturesque enterprise, which was the life-breath of all Lady Morgan's Irish novels. Be she a good schoolmistress or not, on the present occasion Mrs Hall is attractive as a romancer and sound as a moralist.[64]

There is a slight criticism of her dialogue – too flowery and rhetorical – but she 'deserves Ireland's thanks'.

The critic in the *Gentleman's Magazine* was impressed by Mrs Hall's knowledge of the Irish people and her interest in their welfare, and compared her 'lively, spirited and accurate descriptions' to those of Miss Edgeworth but he was unhappy with the 'morbid liberalism' in the story.[65] The reviewer in *John Bull* did not, as we have seen in Chapter One of this study, approve of Mrs Hall's aims – those of 'inculcating political truths, and exhibiting the consequences of a faulty legislation in what regards "justice to Ireland"' – but found her 'far superior to the general run of Irish story tellers'.[66]

The *Atlas* also commended Mrs Hall's skill in story-telling and her power 'to enlist the reader's sympathies – an art which all writers should endeavour to accomplish when either an ethical or a political object is in view'. It was particularly useful in this instance, when the story was designed to expose the evils of absenteeism.[67]

The most perceptive review of *The Whiteboy* was that written by Sir Samuel Ferguson in the *Dublin University Magazine*. In a long article associating Mrs Hall's name with that of William Carleton under the title of 'Didactic Irish Novelists', didactic tales were described as 'a vicious kind of literature', one-sided, prejudiced and artificial, and could be divided into those, which like Disraeli's, tried to convert others to the author's point of view, and those which flattered existing prejudices. Mrs Hall's novel, in common with three new works by William Carleton, *Parra Sastha, Art Maguire* and *Rody the Rover*, fell into the latter category. It pandered to the tastes of those well-meaning English people who believed that Irish people were incapable of helping themselves, but, then: 'Mrs Hall's genius never aspired beyond catering gracefully for pre-existing tastes'. Ferguson, in an outburst of irritated Nationalist feeling, went on:

> The people of England cannot afford to have their ignorance of this part of the United Kingdom increased by representations, however flattering, of their own superiority; for vanity and ignorance have ever been the parents of weakness; and the people of Ireland are much too intelligent not to see, and much too spirited not to resent any disposition to humiliate them for the aggrandisement of self esteem of those whom they daily find to be no more than their equals in any of the pursuits of life, whether practical or intellectual.[68]

Ferguson returns to this theme later in the article, when, having summarised the plot of *The Whiteboy*, he muses on the English belief that Irish people are 'by some physiological necessity so much their inferiors, that, if left to ourselves, we never could emerge from some obscurely surmised depth of barbarism'. This is so far from being the case that 'the Irish people, throughout all parts of the country, and in all the pursuits of life, are prospering rapidly by their own exertions, thinking and acting for themselves, and in many of the highest intellectual pursuits, bidding fair to be the leading people of the United Kingdom'.

Unlike Mrs Hall, Ferguson does not see Ireland in an inferior position *vis-à-vis* England, but as a partner of equal status in a kingdom that is united for mutual benefit. Obviously he does not disagree with her on the necessity and desirability of union, but he does object to the vision of a strong nation taking total command of the affairs of a weaker one. Ferguson considered that the proposals Mrs Hall put forward in her novel for the improvement of the country were so inadequate as to be laughable – 'such great benefits conferred by means so simple as the importation of an English gentleman of no particular ability' – and on the sheer practical difficulties of putting such proposals into action she remained silent. Her wise clergyman, Dean Graves, who gave the young Englishman such good advice, *did* nothing himself, and if a fictional character was unable to deal with the problems set in front of him by the author, how much more difficult it was to solve those problems in real life. Both artistically and politically the book was a failure; the 'inane generalities and commonplaces of mere liberal sentiment' do no good politically and spoil the book as a fiction. The *Dublin University Magazine* had in the past printed admiring comments on Mrs Hall's works; she had been the subject of one of their Portrait Gallery sketches, and had been an occasional contributor to the magazine. Never before had her attitude to Ireland and Irish people been criticized, yet what she had said in *The Whiteboy* was, in essence, no different from what she had often said before. *She* had not changed but the attitude towards her had, in this instance, changed completely. It is in the context of a political novel that Mrs Hall drew such censure on herself, and that only in an Irish Tory Magazine. As we have seen, the English reviewers were ready to accept her as qualified to make political pronouncements, and so was the O'Connellite paper, the *Pilot*, but Ferguson, who correctly read the novel as a plea for the abandonment of Irish political responsibility, would not grant her statements any validity and dismissed her efforts with the comment that *The Whiteboy* 'would not add to her renown for political ability'.

Ferguson, in fact, had the final Irish critical word on Mrs Hall's work. Never again was she taken seriously in Ireland, although she wrote two further novels with an Irish setting – *Midsummer Eve* in 1847 and *The Fight for Faith* in 1869. *The Whiteboy* is her Irish swan song and in spite of some well-deserved contemporary criticism, it is a work to be taken seriously. The novel is well constructed – no longer open to the earlier complaint that her novels were 'painfully spun out to fill three volumes according to the ruthless statutes of Great Marlborough or New Burlington Street'[69] – with a fair amount of dramatic tension, much humour, fine descriptive writing, and an appreciable depth of characterisation. It contains a greater range of vocabulary than is usual in her prose and she uses simile and metaphor with confidence and to excellent effect. There are many lessons in the book but they are so coherently expressed and inserted so effortlessly into the narrative, through the words of the characters, that they are never obtrusive. Mrs Hall is a teacher and her didacticism is seen at its best in this book where she is at ease dealing with the large problems that plagued Ireland – political mismanagement, religious intolerance, legal and illegal violence and poor landlord/tenant relations. Not only by her own lights is the book a success, it may fairly be called one by ours.

NINE

Three Novelists with a common cause

In 1845, the year Mrs Hall's *The Whiteboy* was published, three other novels by Irish authors on the theme of landlord/tenant relations appeared in print. One, *Valentine M'Clutchy*[1], was by William Carleton, the other two, *St Patrick's Eve*[2] and *The O'Donoghue*[3], were by Charles Lever. In Carleton's case, it was no surprise to readers to learn that his new novel was an attack on absenteeism and wicked agents, for these were subjects he had already written about, as had Mrs Hall, but the choice of these themes by Charles Lever was a surprise. He was known as the author of novels of military life – *Harry Lorrequer, Tom Burke of "Ours"*, and *Jack Hinton* – that were fast-paced, full of action, and lightened by humorous anecdotes. Yet there were hints in all these novels of an underlying seriousness and an appreciation of the fact that there were grave problems in Irish society, and they should be read in conjunction with Mrs Hall's book.

There are many similarities between Mrs Hall's *The Whiteboy* and Lever's *The O'Donoghue*. Both novels are set in the same region of Cork and both depend on foreign armed intervention for narrative purposes. In *The O'Donoghue*, it is the French expedition of 1796, while in *The Whiteboy* it is a fictional foray in the year 1822. The plots of the two novels are otherwise very different, but each author shows a preoccupation with the same central problem – the relationship of a landlord with his tenants. In Mrs Hall's novel the young Englishman does everything right and finally wins his tenants' trust and confidence. In Lever's story the English landlord, Sir Marmaduke, makes the classic mistake that Mrs Hall had so often warned against in her writings – he expects his tenantry to behave as they do in England and treats them accordingly. As a result he is easily hoodwinked by 'the idle, the lazy, the profligate' among the people. The more respectable element who knew that his plans were unsuitable to their circumstances, held back, and had nothing to do with them. The others accepted all that they were given:

Cottages were built, rents abated, improved stock introduced, and a hundred devices organised to make them an example for all imitators.

Unhappily the conditions of the contract were misconceived; the people believed that all the landlord required was a patient endurance of his benevolence, they never reckoned on any reciprocity in duty; they never dreamed that a Swiss cabin cannot be left to the fortunes of a mud cabin; that stagnant pools before the door, weed-grown fields and broken fences, harmonize ill with rural pailings, drill cultivation, and trim hedges. They took all they could get, but assuredly they never understood the obligation of repayment.[4]

Every morning now 'the converts to the landlord's manifold theories of improved agriculture' greet him as he appears at his window. They bless him and congratulate him and thank him for his goodness but each and every one has a complaint; the slates have fallen from the new roof because the harrow was put up there to keep them in place, the tenant who was set up as a shopkeeper was resigning because nobody had money to buy his goods, and the man who had been given a hive of bees had had them attacked by the pig and so forth until Sir Marmaduke finished the session in a state of 'stifled rage', muttering about barbarism.

As in *The Whiteboy* there is a wicked agent in *The O'Donoghue*. Called Hemsworth, he has had full control of the estate because Sir Marmaduke, the landlord, is an Englishman who resides on his English estates, and has no interest in his Irish property. Hemsworth, so, has had no problem in extending his own holdings and in amassing wealth. He had

> played his game like a master; for while obtaining the smallest amount of rental for his chief, he exacted the most onerous and impoverishing terms from the people. Thus diminishing the apparent value of the property, he hoped one day to be able to purchase and at the same time preparing it for becoming a lucrative and valuable possession; for although the rents were nominally low, the amount of fees and "duty – labour" were enormous.

It is a very succinct account of the activities of a dishonest agent, and Lever goes on to explain exactly what is meant by 'duty-labour' and the grave consequences it has for the tenant. Because of it the poor man is called

> from the humble care of his own farm to come, with his whole house and toil upon the rich man's fields, the requital for which is some poor grace of a week's or a month's forebearance ere he be called on for that rent these exactions are preventing him from earning ... Duty labour is the type of a slavery that hardens the heart, by extinguishing all hope, and uprooting

every feeling of self-confidence and reliance till, in abject and degraded misery, the wretched man grows reckless of his life, while his vengeance yearns for that of his taskmaster.[5]

The worst feature of duty-labour, as Lever sees it, is the false picture it gives of the relative industry of the agent and the tenant. Those who do not know, the outsiders, the Sir Marmadukes who come to visit a strange land, look upon the agent's well-tended farm, note the contrast between it and the neglected fields of the tenantry, and are ready to believe, having seen the evidence with their own eyes, that the Irish are ignorant and obstinate and are incapable of learning, either by precept or example. The duplicity of agents such as Hemsworth, had, in Lever's phrase, 'drawn a veil between the landlord and the people which no acuteness on either side could penetrate'. The landlord was 'taught to regard the tenant as incurably sunk in barbarism, ignorance and superstition', an easy lesson to teach as it confirmed his deeply-ingrained prejudices, and the tenant saw him as a 'cruel, unfeeling taskmaster'. Hemsworth, when in danger of being discovered, wrote to Sir Marmaduke's son, warning him of the dangers facing his father should he reside on his Irish estate. He would be dealing with a people 'whose demands no benevolence could satisfy, whose expectations no concessions could content'. Their cruelty and savagery when disappointed were fearsome, and he cited actual instances of outrages against landlords. This playing on the deep-seated fears and perpetual sense of danger felt by colonists was effective and the son travelled to Ireland 'armed to the teeth, and prepared to do valiant battle, if need be, against the "Irish rogues and rapparees"' and ready to rescue his father and sister.

The plot of William Carleton's attack on absentee landlords, *Valentine M'Clutchy*, hinges upon the dismissal of an honest agent, Hickman, and his replacement by the vicious and corrupt M'Clutchy by the landlord, Lord Cumber, in an effort to squeeze more money from his unfortunate tenants. Lord Cumber is an absentee who lives in London, runs up huge gambling debts, maintains a yacht, keeps a racing stable and spends lavishly to satisfy the whims of his mistress. The dismissed agent, Hickman, rebukes him for his licentious ways and for his 'indefensible conduct' as a landlord. He reminds him that the tenantry are his best friends and 'his only patrons' those who are 'his chief benefactors, who prop his influence, maintain his rank and support his authority'. He also reminds him of the duties of the landlord to his tenantry and appeals to him to remember that these humble beings have hearts and souls and are 'moved by all those general laws and principles of life and nature, which go to make up social and individual happiness'.[6] Even more soberly, he advises the landlord not to drive an oppressed people into a revolution which would overturn the whole social system. This is Carleton, using Hickman's voice to address the landlords of Ireland, and the absentees for whom he had said in his Preface this book was written to exhibit a 'useful moral'. Valentine M'Clutchy, the villain of the book (or, rather, one of them) exhibits all the faults of which a man is capable. His picture, like that of Abel Richards, the wicked agent in *The Whiteboy*, is too overdrawn to be effective, but there are some fine passages on the damage wrought by a corrupt agent. Again and again Carleton stresses the point that he is cheating both landlord and tenant and ruining not only men but the land itself through bad management. M'Clutchy is more ambitious than many agents because he desires political power and many of the evictions are those of tenants who refuse to support him or his nominee at an election. He is more than willing to create new Forty-Shilling Freeholders for his landlord's political purposes and gives a cynical recipe for their creation, which, though out of character with his personality is, taken on its own, a biting piece of political irony.

Carleton's honest agent, Hickman, is a more likeable creation than the conscientious agent, Lucas, drawn by Charles Lever. Lucas, however, is more believable than the goodly Hickman, or the unctuous Abel Richards. He is an employee who looked after his master's business and had no time for sentiment. Lever did not present him as a monster, more as an efficient machine, unmoved by personal problems. The scene in his office in the square in Galway is vividly brought to life, especially in one little poignant detail where those 'who were deficient in a few shillings, were sent back again, and appeared with the money still in their hands, which they counted over and over, as if hoping to make it more'.[7] Tenants who had always been prompt in their payments got no satisfactory answer when they asked for a renewal of their leases and those who, through 'sickness and distress', had neglected their farms were 'severely reproached'. A total lack of humanity is what is being portrayed here, with no hint of the mutual bond which was meant to unite landlord and tenant. Lucas is merely doing his duty when he refuses to believe that Owen Connor holds his farm rent-free because of an act of generosity on the part of his landlord. The landlord had failed to put this in writing, and now that he is dead, his son, an absentee living in London, has no knowledge of the gift. Lucas insists that the arrears be paid, and Owen is ruined. The original landlord was culpable because he was not sufficiently interested in his tenant's life to secure its comfort, and his son carried this indifference even further by abandoning his estate altogether and leaving its management to an agent. Lever condemns this abdication of landlord responsibility in a direct address to the absentees, when he reminds them of the old attachment of the peasantry to their landlords: 'Such, O landlords of Ireland! it might still have been, if you had not deserted the people ... Your presence in the poor man's cabin – your kind word to him on the highway – your aid in sickness – your counsel in trouble, were ties which bound him more closely to your interest, and made him more surely yours, than all the parchments of your attorney, or all the papers of your agent'.[8] Later he cries out to them: 'You left them [the tenantry] to drift on the waters, and you may now be shipwrecked among the floating fragments'.

This warning of disaster to come shows Lever as a man concerned for the cause of the landlords as well as that of the tenants. Indeed, he prefaces it with the statement that he is 'not joining in the popular cry against the landlords of Ireland'. Some of the book's reviewers found this an extraordinary statement, but their reading must have been rather superficial as he makes it very clear that it is the absentees he is criticising and not the residents. Indeed, the whole story depends on the fact that the landlord *is* an absentee – if he had been a resident Owen would have been able to get help from him. As it is Owen travels to London to tell the landlord his sad tale and to re-state his claim. This journey is akin to the one made by Kate Connor in Mrs Hall's first book of Irish sketches, when she travelled to London to seek the help of her landlord, and is also similar to that made by Carleton's Owen McCarthy in the story 'Tubber Derg' who went to Dublin for the same reason. Kate Connor's mission was a success, but Owen McCarthy's was a failure. Carleton had sometimes subscribed in his fictions to the romantic idea of the landlord who will right the peasant's wrongs if his subordinates will allow him to learn of them but he subverted the whole notion in Owen's story 'Tubber Derg' because the landlord refuses to deal with him and insists that he take the matter up with the agent. Owen goes home a beaten man, his last and best hope gone. Lever's Owen Connor has no luck in London either, but this is because the landlord is abroad, and eventually, by a stroke of good fortune, Owen encounters him on the estate in Galway and all is happily resolved, with the young landlord vowing to reside on his estates in future and to see to the well-being of his tenants.

Carleton also brings the story of *Valentine M'Clutchy* to a satisfactory conclusion: the wicked are punished and the virtuous are rewarded. Even more importantly, the weak are protected and are guaranteed future security, for on the death of the avaricious landlord his kindly and conscientious brother takes over the running of the estate, promising to live there and to see to the welfare of the tenants. In both of these novels the message is loud and clear – absenteeism is a dereliction of moral duty, and those who sin in this way will suffer just as surely as their victims will suffer. There will be practical and immediate consequences of such neglect, and both Carleton and Lever spell these out for their readers. Through Hickman, the honest agent in *Valentine M'Clutchy*, Carleton reminds the landlord Lord Cumber, that the tenantry are his best friends and warns them, as does Lever in *St Patrick's Eve*, that the social system on which they depend may be wrecked by their sins of omission and commission. He makes an even clearer statement at the end of the book, after Lord Cumber has been killed:

> Strange indeed, it is, that men placed as his Lordship was, should forget a principle, which a neglect of their duties may one day teach them to their cost; that principle is the equal right of every man to the soil which God has created *for all*. The laws of agrarian property are the laws of a class; and it is not too much to say, that if the rights of this class to legislate for *their own interests* were severely investigated, it might appear, upon just and rational principles, that the landlord is nothing more nor less than a pensioner upon popular credulity, and lives upon a fundamental error in society, created by the class to which he belongs.[9]

Speaking directly to landlords Carleton added: 'Think of this, gentlemen, and pay attention to your duties'. This passage which bluntly states that the claim to property is a very tenuous one and is not based on any ancient or immutable law is as close as Carleton (or any of his fellow Irish writers at that time) came to suggesting an overthrow of the existing class system.

A modern commentator has spoken of the 'Marxist indignation, not Tchehovian melancholy',[10] that pervades *Valentine M'Clutchy* but contemporary critics did not notice any revolutionary tendencies in the work. The *Atlas for India* noted that Carleton's design 'to exhibit the crimes perpetrated by agents in the absence of their employers or, as was too frequently the case, with their approbation, is fully accomplished',[11] and the *Spectator* referred to the 'knavish agents who took advantage of the ignorance and apathy of spendthrift absentee landlords to wring from the impoverished tenantry the last shilling',[12] but the book was not generally well received because it was felt to be too partisan and too coarse. Lever, too, sounds the alarm for landlords. By their failure to do their duty as owners of property, and natural leaders of the peasantry, they are undermining the foundations of their own power and allowing their tenantry to fall under the influence of agitators. This is his diagnosis of the ills of Ireland, and crimes of violence, burnings and assassinations are the result of 'social disorganisation', rather than political grievances. Lever is well aware that his work is not likely to be read by peasants and it is directed specifically at landlords in the hope that their hearts may be touched by the sad story of 'one humble man' but that it may have an even stronger appeal to their heads and that they will see that it is in their own interests to assume leadership of their tenants and so avert catastrophe. If they had done so in the past 'the tares would have been weeded from the wheat; the evil influence of bad men would not have been suffered to spread its contagion throughout the land'.[13]

It was obvious to reviewers that this was a tale with a moral, and they applauded Lever's purpose, stated in the Preface, to demonstrate that 'prosperity has as many duties as adversity has sorrows'. The *Literary Gazette*, however, fearful that readers might be expecting a doleful sermon, pointed out that the quoted passage 'must not be taken as a sample of the incidents of the tale, and striking pictures of natural scenery, national manners and peculiar feelings'.[14] In other words, the story may be read 'with equal pleasure and profit'. There was general agreement among the literary critics about the artistic merits of the novel: 'an exquisite little tale', 'graphic delineation of character' 'engaging and captivating style', 'well-constructed', 'vivid and life-like reality' are some of the phrases used, but the main attention was focussed on the moral purpose of the novel. All the reviewers agreed that in it Lever had exposed the evils of absenteeism (though the *Atlas* critic wondered if Carleton's *Valentine M'Clutchy*, published a few weeks earlier had inspired *St Patrick's Eve*[15]) and that he was right to do so. It might even succeed in its purpose of awakening the absentee landlords to a sense of their duties, and *The Sunday Times* stated: 'In it the Irish landlord will see a more correct and touching picture of the evils of the absentee system than he could ever gather from treatises of the most formal and elaborate kind'.[16] The *Athenaeum*, however, while granting that 'the blessings of local residence are painted in forceful colours' feared that 'the absentee landlords of Ireland were not to be reached by a fiction'. 'Besides', the writer added, 'are they not also the victims of a system?'[17] He did not expand on what that system might be, nor how it might be changed, but leaves the statement dangling in mid-air.

The *Spectator* took a harder look at the story and its moral purpose and concluded:

> As a didactic story it fails, by taking exceptions for its instances. Owen Connor is an exception; the landlord is an exception; and the agent is an exception – we trust, not that he is a raw-head-and-bloody bones man of the old school, but too indifferent to the commonest sense of right, according to English notions. Neither are Mr Lever's positions unassailable; and he certainly falls into a common error of Young Ireland, by making his countrymen out to be too helpless. Owen Connor has his farm rent free; but he does not make much of this advantage because his landlord is not there to look at him and pet him up. Surely this endless desire to lean on others, to look for aid other than from the only source of aid, innate energy and self-exertion, is too much of an Irish failing to be inculcated in a didactic tale.[18]

The stereotyped colonial view of the lazy Irish peasant is very obvious in this comment, made ludicrous by the contradiction it contains – if Owen is an exception, how then can he be suffering from a typically Irish fault? Furthermore, it is proof that the critic did not read the book properly; Owen did not fail to make anything of his farm but lived in tolerably comfortable circumstances until he was faced by the agent's refusal to believe that he did indeed have his land rent-free and his insistence that Owen pay arrears. In Ireland, the *Nation*, at one time fiercely critical of Lever, declared with grace and generosity that 'the scales have fallen from the eyes of Mr Lever. Bolder and sounder views of the tenant question we have seldom met'.[19]

Lever's other novel of 1845, *The O'Donoghue*, was a more complex piece of work than *St Patrick's Eve* and while it is concerned with the problems of landlords and tenants it does not, unlike *The Whiteboy* which it most resembles, or *St Patrick's Eve* and *Valentine M'Clutchy*, end happily with a landlord who resides on his Irish estates and gives moral and practical guidance to his tenants. It is, in fact, the story of the collapse, both financially and morally, of an ancient Irish family – the O'Donoghues of the title. What were accounted virtues in Irish society – hospitality, generosity, good-living – were the vices that destroyed them, and shattered their authority over their tenants. Wealth and power both disappeared, and it was the English newcomers who took on the responsibility of caring for the tenants. The new landlords, however, did not understand the peasant mentality, and although they tried to improve the lot of the peasants and were kindly and charitable towards them, they were flattered and fooled by cunning natives who took all they could from them, but gave nothing in return. Lever presents this as a metaphor of the state of Ireland, abandoned by its native rulers, leaderless, and subject to the rule of outsiders, who are unequal to the task. The honest men, those of integrity, rebel against the outsiders, but cannot succeed and must go into exile. No happy conclusion is reached in the book, nothing is resolved and the main characters abandon the country. The Irish aristocracy has collapsed under the weight of its own extravagance and improvidence, rebellion has failed, there is no help from other countries, so English rule, misguided as it is, must continue. The tenants are no better off than they ever were, because the O'Donoghue estate is tied up in the courts and ownership is difficult to prove. Almost the last words of the novel are spoken by an old Irish beggarwoman who answers questions put to her by travellers curious to know the history of the valley they are visiting. She speaks with bitterness of the former inhabitants, Irish and English alike, tells how their houses are now the haunts of the crows and adds: '"It is little worth while remembering them ... if you want to pity any one, pity the poor, that's houseless and friendless"'.[20] Lever is more censorious and critical of the ancient Irish families than is Mrs Hall in *The Whiteboy*. Lawrence MaCarthy, the last of his line, is landless because the MaCarthys were dispossessed by the invader, and this is the reason why he becomes involved with outlaws and resorts to violence. He is not a noble figure, but he has remnants of grace and his death is not a shameful one.

In *The Whiteboy* Mrs Hall made clear her views on those who agitated for the repeal of the Act of Union. She had nothing but contempt for them and believed them to be acting against the best interests of their country. No such clarity informs Charles Lever's novel *The O'Donoghue*. The story is set in 1796, so the action takes place in pre-Union days and Lever writes nostalgically of what he saw as a golden time in the life of Dublin, then a capital city. 'Provincialism had not then settled down upon the city', he says, and 'the character of a metropolis was upheld by a splendid court, a resident Parliament, a great and titled aristocracy.' The 'tone of society had all the charms of a politeness now bygone', and conversation then did not consist of the 'easy platitudes of the present day'. However, it is not the author's business to seek reasons for this change:

It is not our duty, still less our inclination, to inquire why have all the goodly attractions left us, nor wherefore is it, that, like the art of staining glass, social agreeability should be lost for ever. So it would seem, however; we have fallen upon tiresome times, and he who is old enough to remember pleasanter ones has the sad solace of knowing that he has seen the last of them.[21]

Nostalgia and a certain amount of sad resignation is all that is expressed. There is no clear vision of what the future might be for Ireland, none of Mrs Hall's buoyant optimism. Taken in this sense the work is less of a political novel than is Mrs Hall's *The Whiteboy* although she was adamant that she was never concerned with politics. Lever went on, of course, to write a novel that was truly political, *Lord Kilgobbin*, in 1872, but Mrs Hall made no such literary or political development. Her last Irish novel, *The Fight of Faith* (1869) is a crude piece of religious propaganda set in the seventeenth century and is a regression rather than a progression. Charles Lever's other novel of 1845, *St Patrick's Eve*, the last he wrote while still resident in Ireland, is not political except in the sense that the subject is political, and that landlords still either hold political power, or influenced those who did, but it does trace the roots of violence back to the injustices inflicted upon the Irish peasantry. Reviewers generally were rather puzzled by Lever's apparent change from the jolly *persona* of 'Harry Lorrequer', the supposed writer of military yarns to that of a serious social commentator. The *Nation*, however, which had been Lever's fiercest critic, and had employed Carleton's pen to scarify him, now hailed his 'growing love of Ireland' and pronounced *St Patrick's Eve* 'a political novel as much as THE NATION is a political journal'.[22]

Unlike Mrs Hall and Charles Lever, William Carleton made it clear that his novel of 1845, *Valentine M'Clutchy*, was a political one. It was written, he says in the Preface, not only to bring landlords and agents to a sense of their duty, but 'to teach the violent and bigoted Conservative – or, in other words, the man who *still* inherits the Orange sentiments of past times – a lesson that he ought not to forget'.[23] M'Clutchy, the wicked agent, is 'an Orange Tory', and can, therefore, expect political advancement, because, as a character in the novel says, '"there is no such successful recommendation as this violent party spirit even to situations of the very lowest class. The highest are generally held by Orangemen, and it is attachment to their system that constitutes the only passport nowadays to every office in the country, from the secretary to the scavenger"'.[24] Carleton gives much information about the organisation of the Order, describes in detail the meeting of an Orange Lodge, and recounts some of the excesses of the members, but insists that the worst feature of Orangeism is its assumption of those powers which should belong to the legitimate government of the country. Real political power is in the hands of the Order, which is identified with the Protestant church and which manipulates those politicians who sit in Parliament.

Carleton traces the decline of legitimate political activity through corruption, neglect, and the lust for personal power back to the Union, and although the author does not intend

> to discuss the merits of either the Union or its Repeal, but in justice to truth and honour, or perhaps, we should rather say, to fraud and profligacy, we are constrained to admit, that there is not to be found in the annals of all history, any political negotiation based upon such rank and festering corruption, as was the Legislative Union. Had the motives which activated the English Government towards this country been pure and influenced by principles of equality and common justice, they would never have had recourse to such unparalleled profligacy.[25]

Old Tom Topertoe, the local landlord, had voted for the Union, and was rewarded with the title of Lord Castle Cumber. He was, says the author, typical of his times, and 'of shame or moral sanction he knew nothing', so that he was 'not only the very man to sell his country, but to sell it at the highest price, and be afterwards the first to laugh, as he did at his own corruption, and say that '"he was devilish glad he had a country to sell"'. With sellers such as Topertoe, and buyers such as the English politicians, no wonder the Union was passed, and no wonder that no good could come of it. With English motives under suspicion, the whole apparatus of English government in Ireland is defective. Significantly, Valentine M'Clutchy's house is named Constitution Cottage, and the corruption of which it is the local centre is a microcosm of the greater corruption.

Carleton's objections here to the Union are based on a revulsion against the methods by which it was achieved, and he sees the failure of legitimate political activity and its repression by the Orange Order as stemming from a lack of ethics. The 'good' agent in the novel, Mr Hickman, wonders whether he can 'as an honest man ... render political support to anyone who had participated in [the Union's] corruption or recognised the justice of those principles on which it had been accomplished'.[26] Hickman loses his position as agent to Lord Cumber partly as a result of these political beliefs, because the absent landlord needs a man '"on his property who is staunch"'. An agent was expected to marshall the votes of tenants at election time and ensure that they voted for their landlord, or their landlord's candidate, and an agent opposed to Unionist principles, or rejecting them utterly, was a dangerous liability. Hickman, then, is the portrait of the honest man of affairs who is badly needed in Ireland, but who has no place in an unjust system. While Carleton is eloquent on the subject of the ills arising from the Union, he does not suggest any political alternative. Nowhere in the book does he, in his authorial role, or through the voices of his characters, support a substitute legislative assembly, not even a devolved one, for Ireland. Given his denunciations of landlords and clergymen, who might have been expected to form such an assembly, this is understandable. The story ends without any significant change in the social and political system; landlords are still in charge of their estates, although a resident takes the place of an absentee, and the Orange Order is still in control of the political system. Most of the reviewers of *Valentine M'Clutchy* accepted the novel as an exposé of bad landlords and their agents and commended the pathos with which the author invested his descriptions of their victims' sufferings, but some reviewers detected political didacticism in the work. The critic in the *Athenaeum* believed that one of the purposes of the tale was 'to aid the cause of Repeal', but asked if Carleton wished 'to restore the Parliament that was so notoriously venal'.[27] A review in *Tait's Edinburgh Magazine* touched on the same question, but noted that he left 'the reader to gather his opinions in the great question of Repeal, from violent vituperation of the Union, and the corrupt instruments by which it was effected'.[28]

Religion in Ireland was not a subject that contemporary novelists could ignore, and Mrs Hall, William Carleton and Charles Lever all dealt with the subject in their own way. In *The Whiteboy* Mrs Hall preached the virtues of religious tolerance and deprecated fanaticism and bigotry, but her emphasis was more on the personal and private than on the public and social. Carleton, on the other hand, boldly tackled the whole question of religion and society in *Valentine M'Clutchy*. In the novel Carleton depicts a society where all the power is in the hands of Protestants, as the agent, M'Clutchy, points out, while without exception – landlord, agent and attorney – they abuse it, and the sufferers are the Catholics. Religion for these men is a social force, not a spiritual one, and Carleton points the contrast between them and Catholics, especially in his description of Mary O'Regan's death. She was a poor Catholic woman, whose husband and sons had died because of the agent M'Clutchy's cruelty. She lost her reason but recovered it at the point of her own death. When she came to her senses she found herself alone in a churchyard in the mountains where the bodies of her family lay in their graves. Her prayer to God is suffused with faith and a sense of resignation:

"Merciful Father", she ... exclaimed, "do not – oh! do not suffer me to die in this wild mountain side, far from the face or voice of a human being! There is nothing too powerful for your hand, or beyond your strength or your mercy, to them that put their humble trust in you. Save me, ah God, from this frightful and lonely death, and do not let me perish here without the consolations of religion! But if it's thy blessed and holy will to let me do so then it is my duty to submit! Give me strength, then, to bow to thy will, and to receive with faith and thanksgivin' whatever you choose to bestow upon me! And, above all things, O lord, grant me a repentant heart, and that my bleak and lonely death-bed may have the light of glory upon it! Grant me this, O God, and I will die happy even here; for where your blessed presence is, there can be nothing wantin'"![29]

The prayer is answered and the priest arrives in time to give her the last rites. Her dying words are of forgiveness for those who have wronged her and she 'passed away into the happiness of God's love, which no doubt, diffused its radiance through her spirit that was now made perfect'. The priest, Father Roche, had been educated on the Continent and is 'a scholar and divine', with the sheen of his foreign training still gleaming upon him; 'his manners with all their simplicity are those of a gentleman, possessing as they do all the ease, and when he chooses, the elegance of a man who has moved in high and polished society'.[30] (In this he is like Mrs Hall's priests from an older world, and is unlike the new Maynooth-trained clergy.) Significantly, Carleton shows the priest dissuading a band of Ribbonmen from a murder, by the force of the moral argument he uses. This may be read not only as a specific rebuttal of the charge that priests condoned violence (and as marking a change in the author's own attitude to Catholic priests) but also as Carleton's appeal to his countrymen to remember that justice belonged to Almighty God alone.

The deathbed of Squire Deaker, the licentious and profligate father of Valentine M'Clutchy, is very different from that of Mary O'Regan. His end is hastened by over-indulgence in drink and sexual activity and the best that can be said of him by the onlookers is that he kept his vow to die whistling the Orange tune 'The Boyne Water'. Unlike Mary O'Regan he speaks no words of faith, nor does anyone else in the room. The servant, Lanty, talks about black ravens in the beech trees waiting with foreknowledge of Deaker's death and the attorney, Solomon M'Slime, opines that Satan now has a companion, but otherwise, there is no mention of anything that is not material or temporal. Carleton's authorial comment emphasises the lack of spirituality when he says that Dealer's 'political Protestantism ... regulated his life, but failed to control his morals'. A parallel to Mrs Hall's attack upon the proselytising Abel Richards, agent and Presbyterian, is seen in Carleton's creation of Solomon M'Slime, the hypocritical attorney who uses Evangelicalism as a means of self-advancement and self-indulgence. He is no more believable than is Abel Richards but through his character Carleton is able to attack the religious intolerance of the Evangelical movement as viciously as he had once attacked the priests of the Roman Catholic Church. In his Preface to *Valentine M'Clutchy* he had publicly regretted his former strictures though without actually mentioning that Church by name, saying:

> A more enlarged knowledge of life, and a more matured intercourse with society, have enabled me to overcome many absurd prejudices with which I was impued. Without compromising, however, the *truth* or *integrity* of any portion of my writings, I am willing to admit, which I do frankly and without hesitation, that I published in my early works passages which were not calculated to do any earthly good; but on the contrary, to give unnecessary offence to a great number of my countrymen.[31]

Tait's Edinburgh Magazine in its review of the novel commended Carleton for 'the frank apology made for many offensive passages against the Catholic party made in his previous works' and noted that he was 'no longer the satirist or adversary of the Roman Catholic priesthood', but that 'what is called the Evangelical has with him taken the place of the Ultra-Catholic party'.[32]

The novel is an examination of sectarianism as well as an exposé of Protestantism as a political force, but an even balance is not held between Catholicism and Protestantism. There is more criticism of the Established Church than there is of the Roman, and in the exchange between Mr Lucre, the Protestant minister, and Father M'Cabe, the Roman Catholic curate, the most forceful arguments are put in the mouth of the curate. In reply to the minister's taunt that the Roman Catholic church is 'a mockery of religion', Father M'Cabe says to him

> "I cannot but remember the mockery of religion presented by your proud and bloated bishops who roll in wealth, indolence and sensuality; robbing the poor while they go to h – l worth hundreds of thousands. I cannot forget that your church is a market titled and venal for slaves, who are bought by the minister of the day to uphold his party – that it is a carcase thrown to the wolfish sons and brothers of the English and Irish aristocracy – and that its

bishops and dignitaries exceed in pride, violence of temper, and insolence of deportment, any other class in society."[33]

The Rev. Phineas Lucre personifies the worst features of the Established Church as Carleton saw them; he is devoid of any good qualities and is hypocritical, lazy, and 'without piety to God, or charity to man'. His political connexions and consequent preferment in the church are stressed, and Carleton accuses him of spying for the Government. By comparison, Father Roche, the Roman Catholic priest, is shown as a compendium of the virtues and it is he who calls for an end to sectarianism and bigotry, asking: '"When will this mad spirit of discord between Christmas ... be banished by mutual charity and true liberty from our unhappy country"'?[34] The *Atlas for India* noted the imbalance in the novel and said that Carleton

> paints almost all his Catholic characters as virtuous, and noble, and humane while the generality of the Protestant personages are ferocious and cowardly despots. It is true the author attempts to be fair in his delineation of the various personages who figure in his groups, and he casts vices and profligacy with a liberal hand amongst some of his Roman Catholic characters, as well as his Protestant ones; but the two most villainous of the whole of his *dramatis personae* are ... Orangemen and Protestants, while the most angelic attributes are bestowed upon Father Roche and the family of the M'Loughlins and they are Roman Catholic.[35]

Charles Lever's work contains few references to the sore question of religion in Ireland, and his cool detachment is far removed from Carleton's passionate crusade against the evils of organised religion and Mrs Hall's appeals for an end to intolerance. His cunning Irish peasant, Mickey Free, who appears in *Charles O'Malley* (1841) was 'a devout Catholic, in the same sense that he was enthusiastic about anything; that is, he believed and obeyed exactly as far as suited his own peculiar notions of comfort and happiness'.[36] He is a 'type' (as Lever acknowledges in his Preface to the 1872 edition) and so, too, is Father Loftus, the convivial priest in *Jack Hinton, the Guardsman* (1843). Father Loftus at first seems to be another of the hard-drinking and gambling priests who appeared in Lever's first novel, *Harry Lorrequer* (1837) and whose portrayal gave offence to Catholic readers, but as the story goes on he is revealed as a much more complex character. He is devoted to his priestly duties and is an opponent of political violence. It is in his mouth that Lever puts the speeches that illuminate the character of the Irish peasantry for the young guardsman and he is the means of bringing Hinton to an understanding of some of the problems of Ireland. Under his rather uncouth exterior there is a thoughtful, sensitive and intelligent personality, well-read and well-educated.

Father Rourke in *The O'Donoghue* (1845) is an equally cultured man, and Lever shows him as sharing Father Loftus's distrust of violence. He refuses to rally the peasantry to fight alongside the French expeditionary force that had landed in Bantry, as much from a hatred of bloodshed as from a disinclination to ally himself with the enemies of religion. Lever adds the authorial comment that 'the dreadful wrongs inflicted on the Roman Catholic church during the [French] Revolution could not be forgotten or forgiven by the priesthood'.[37] This reminder by Lever that Roman Catholic priests had a fear of revolutionary violence was well-timed, coming as it did when Irish Nationalism seemed to be escaping from the control of O'Connell and moving towards an expression of physical force. The priests, whose influence was feared by English observers, had been presented by Carleton as ambivalent on the subject of violence, but Lever showed them on the side of restraint. The only mistake, from the point of view of religious propriety, that Lever made in *The O'Donoghue* was the telling of a joke about a priest and a dying man. This drew a stern caution from the English Catholic periodical, the *Tablet*: 'It is a duty to warn Catholic parents not to suffer their children to read a book which ... speaks with the scoffer's tone of sacred things, and treats of an article of faith with horrible impiety'.[38]

Lever's flippancy about religious matters, unusual among Irish writers of the time, would seem to have sprung from not so much a tolerance of religion as an inability to understand why differing dogmas divided people so sharply. In his novels the characters are born into a religion, and so into a way of life, and, consequently, they are not concerned with religious differences as such. Lever was aware of the social connotations of religion in Ireland, as he demonstrated in *The O'Donoghue* when he showed young Herbert, a brilliant student, converting to Protestantism, not from conviction, but because it is the only way in which he can gain scholastic advancement in Trinity College. As a result, not only does he win the coveted academic prizes, but he changes in every way for the better. He becomes happy and self-confident, and time spent in the 'cultivated and polished circles' of Protestantism make him 'a youth of graceful and elegant demeanour'. Consciously or unconsciously, Lever's references to Herbert's change of religion echo what the Catholic periodical the *Dublin Review* had had to say in 1836 about the dangers to Catholic youth that lurked in Trinity College, 'the great nurse of infidelity in Ireland'. These impressionable boys were exposed to heretical opinions, and worse: 'the tenets of Christianity [were] decried as ungentlemanly'.[39] It is arguable, however that Lever did the reading public, in Ireland as well as in England, a service when he presented Irish priests as human beings, with a capacity for laughter and an uncomplicated appreciation of the pleasures of good food, good drink and good company. Jack Hinton's assertion to Father Loftus when they met on the canal boat, '"I'd rather pronounce on your punch than your polemics"' is a fair summation of the message Lever is sending to his readers, and is a sophisticated way of teaching religious tolerance.

The violence of Irish life was realistically conveyed in each of the four novels under review and the reasons for its prevalence were, to greater or lesser extent, examined by the authors. Mrs Hall was aware that much of this violence was connected with the question of land but she also examined the violence that was inherent in the system of government – legitimised violence. Charles Lever, although he did not actually use the military metaphor, saw Irish peasant violence with the eyes of one who was familiar with the chain of command that exists in armies. When the chain of command is broken and the superior officer deserts his post, the men under his command no longer have a final court of appeal, their welfare suffers, and they become so resentful that mutiny is a real threat. The violence of these peasants, the lower ranks, is kept under control only by firm control from their landlords, their officers, who must themselves obey certain rules and must provide inspiration to those below them. Lever's peasants are violent and he contrasts the fighting for fun at the fair in the opening chapter with the real and deadly violence towards the end of the book when dispossessed men murder agents and land-grabbers. His hero, Owen Connor, although driven to the edge of mutiny by injustice, and despairing because his superior and natural protector, the landlord, had deserted the country, draws back at the last moment, sickened by the killings. The landlord returns, his heartless second-in-command is deprived of power, Owen is forgiven (in army terms he would have been shot) and order is restored.

In *The O'Donoghue*, that story of an ancient Irish family in the process of disintegration, Lever emphasises that there are no clear systems of control because everything is breaking down. Neither the Irish landlord nor the new English owner of the estates is capable of leading or guiding his tenants. The younger generation is equally ineffectual and the potential for future violence is very real and very strong. The story is set in 1796, and the French expedition to Bantry Bay forms the climax of the book. Lever's attitude towards Irish rebels has softened, and although he paints a black picture of an Irish informer, and refers disparagingly to those who 'are patriots – because they are paupers', his criticism of the United Irishmen is not of any violent methods they might have used to further their aims but of the party's lack of organisation and cohesion. Young Mark Donoghue is presented as typical of the young men '"of the best families of the country, whose estates are deeply encumbered – heavy mortgages and large dowries weighing them down – are ready to join in any bold attempt which promises a new order of things"'.[40] When he fails to gather enough men in Cork to gather around the French expeditionary force, and leaves Ireland to take up a commission in the French army he is greeted on board the French ship by an officer who remarks: '"Ireland is not ripe for such an enterprise. There's no slavery like dissension"'.[41] This considered comment reflects Lever's belief in order and in military discipline but it also marks the advance he has made towards an appreciation of Irish problems. It is no longer a matter of decrying violent methods of expressing national dissatisfaction, but of sober reflections on the difficulties of moulding a national will and achieving a national purpose. The violence of physical action is beginning to be of less interest to him than the struggle between different sets of ideas which can be just as violent in its assault on traditional customs and modes of thought. Unlike Carleton and Mrs Hall, Charles Lever had never announced that he wished to cure the faults of his countrymen, so where the other two authors examine not only the effects, but the causes of domestic, or personal violence, he is concerned with public and highly visible violence. Some of it is sanctioned by law, as in military action, but he makes it clear that even there it has limits, and he stresses that its results are not always very happy ones. It can also be misdirected and although it is legitimate it may defeat its own ends. The violence practised outside the law is equally reprehensible and equally ineffective as a means of solving political and social problems. These views of Lever's are in tune with those of Carleton and Mrs Hall but Lever goes further than the condemnation of violence expressed by the other two writers either in their authorial voices or through the words of their fictional characters, and hints in *The O'Donoghue* at a wider examination of the whole problem of violence in a colonised nation.

William Carleton, like Mrs Hall, described and decried the violence that was diffused through Irish society and that appeared in many forms – physical, verbal, and sexual. An ominous and pervasive presence in *Valentine M'Clutchy* is that of the Orange Order, an ideology, which like that of the Whiteboys, is perverted for private profit. 'Evory Easel', the so-called artist, (in fact the landlord's brother) writing to him in London, granted that the qualifications for an Orangeman specified in the society's code were admirable. He says in his letter:

> had the other portions of it [the code] been conceived and acted on in the same spirit, Orangeism would have become a very different system from that which, under its name, now influences the principles and inflames the passions of the lower classes of Protestants, and stimulated them, too frequently to violence, and outrage, and persecution itself, under a conviction that they are only discharging their duties by a faithful adherence to its obligations.[42]

Unfortunately, these obligations are no more than recommendations, 'they stand there as a thing to look at and admire, but not as a matter of duty'. If strict observance had been enforced they 'might have raised the practices of the institution from many of the low and gross atrocities which disgraced it'. Of the rest of the code, Easel has no good to say, but sums up by describing the order as 'a body irresponsible and self-constituted, confederated together, and trained in the use of arms, (but literally unknown to the constitution), sitting, without any legal authority, upon the religious opinions of a class that are hateful and obnoxious to them – and, in fact, combining within themselves the united offices of both judge and executioner'. Carleton suggests that this illegal army, dedicated to the repression of Roman Catholics, was uniquely suited to further private purposes, and M'Clutchy, the vicious agent, was quick to use it for evictions, and as an instrument of general terror. M'Clutchy's son, Phil, also used it in furtherance of his scheme to compromise the young girl who had rejected his proposal of marriage, by arranging that the Orange yeomanry should search her father's house for a supposed *cache* of arms, while he, having gained admittance by a ruse, would be found in her bedroom. The scheme worked, and the girl's reputation was severely damaged. This type of occurrence was not uncommon in real life and was accompanied by great violence. The yeomanry, as Carleton described them

> with loaded fire-arms, proceeded, generally in the middle of the night or about day-break, to the residence of the suspected person. The door, if not immediately opened, was broken in – the whole house ransacked – the men frequently beaten severely and the ears of females insulted by the coarsest and most indecent language. These scenes, which in nineteen cases out of twenty, the Orangemen got up to gratify private hatred and malignity, were very frequent.[43]

In *Valentine M'Clutchy* the violence on the part of the Orangemen was described by Carleton as unrestrained, while the opposing party was held back from similar violence by the efforts of a priest. His force of character and moral conviction persuaded them how wrong it was to retaliate and they took no action. Only the Catholic clergy can prevent peasant violence, Carleton is saying, but the task grows daily more difficult when 'law and justice [are] so partially and iniquitously administered as to disorganize society, and make men look on murder as an act of justice, and the shedding of blood as a moral triumph, if not a moral virtue'.[44]

Carleton shows sexual harrassment as one of the expressions of violence on the part of M'Clutchy and his son, and fathers tell the meeting of Ribbonmen how their daughters are seduced and ruined and how they are jeered at when they seek justice: '"First they injure us, and kick us about as they plaise, and then laugh at and insult us"',[45] says one man, recounting how Phil told him that his daughter should be proud to bear Phil a child. Those whose daughters did not yield to Phil were penalised, and eventually evicted from their farms. Phil is shown lusting after a poor and virtuous widow, who at first, in her innocence, does not understand his insinuations, but who, when she does, is offended and humiliated. Carleton leaves us in no doubt that she is a victim without much chance of escape. He speaks of 'poverty and dependence on one side and cold, cruel, insolent authority on the other',[46] and the helplessness of those who are at the mercy of M'Clutchy and his son. It is made tragic by the fact that there is no court of appeal for their decisions, no way in which they can be brought to account for their crimes, because they are supported by the system, and their victims have no hope of justice. Individual violence is reprehensible, but state violence, or legitimised organised violence, is infinitely worse because it destroys the basis on which a civilised society is built.

Carleton perceived that priests were connected in the popular mind of the time with violence, and his portrayal of Father Roche, the priest who restrained the Ribbonmen in *Valentine M'Clutchy* is very important when viewed in that light. In order that the point he is making will not be missed, Carleton underlines the priest's action, emphasising that the wicked agent and his son were saved by 'the very man whom they termed '"a rebellious Popish priest"'.[47] The other priest who is portrayed in the book, Father M'Cabe, is a very different type, for he is a violent man himself, nicknamed M'Flail because of a tendency 'to use the horsewhip as a last resource, especially in cases where reason and the influence of argument failed'.[48] He uses it in an altercation with Darby O'Drive, the

Catholic who had turned Protestant, first in the air to give emphasis to his scolding of the man, and then in earnest on his shoulders when his patience runs out. In Father M'Cabe's confrontation with the Protestant parson, Mr Lucre, his tongue lashes as fiercely as any whip in answer to the parson's insults. The two men have a 'religious discussion', which is not about dogma but about politics and each expresses the prejudices he holds about the other's party, descending eventually to personal abuse. Religion, politics and violence, are the constituents of Irish life, all tragically mingled.

Mrs Hall's *The Whiteboy* was seen by reviewers as didactic, a term of praise in some instances but a derogatory one in the case of the review in the *Dublin University Magazine*, which bracketed the work with Carleton's short novels *Parra Sastha, Roddy the Rover*, and *Art Maguire*, all stories with a lesson to teach. By the same standards applied by most critics *Valentine M'Clutchy, St Patrick's Eve* and *The O'Donoghue* may all fairly be described as didactic – and not in a pejorative sense. Charles Lever is not an author with whom one would associate a teaching purpose but in *St Patrick's Eve* he is overtly didactic in a story which has no purpose other than to recall landlords to a sense of their duty towards their tenants, but his main characters are much more than symbols. The landlord is careless, but not cruel, and it is in keeping with his character that he should be so absorbed in his youthful pleasures that he forgets to honour a generous promise made by his father. The peasant who suffers is shown by Lever to be a great deal more than a simpleton with a feudal attachment to his master and an animal contentment in his familiar home. He is not very wise in the ways of the world, but he is resourceful enough to make his way to London in search of his landlord, convinced that he will help him, and Lever very cleverly presents this peasant as a man capable of making judgments on what he sees on the way. The device of seeing England through Irish eyes is a good one and extremely suitable for didactic purposes. Owen Connor's réflections on what the English call 'poverty' put the Irish problem into perspective in a way that is more effective than a list of comparative statistics could ever be, yet it springs naturally from the events of the story, and the lesson is absorbed almost without our realising it. *St Patrick's Eve*, like Carleton's *Parra Sastha*, is that rare creation, a novel which is expressly didactic in purpose and at times in manner, as when Lever directly addresses absentee landlords, yet it works on all levels and entertains and instructs at the same time.

The *O'Donoghue* is not obviously didactic, but it teaches lessons nonetheless – the evils of absenteeism, the futility of violence, and the terrible effects of moral failure. Lever's didacticism, like that of Mrs Hall in *The Whiteboy* is handled skilfully, and is not obtrusive, but in Carleton's case the lessons leap off the page. It is a major fault in such a writer. He has the rare gift of creating an entire self-contained world within the covers of a book, a world full of real people, with whom we laugh or cry or rage at injustice, but he seems unable to leave that world to exist on its own. He insists on pointing out its features to us and on telling us to learn from them. It is not enough that every story of his implicitly or explicitly teaches a lesson, he interjects authorial comments, harangues the reader, addresses those he sees as villains, and even quotes facts and figures to make his case. It is significant that three such disparate writers as Mrs Hall, William Carleton and Charles Lever should have, in the same year, published novels with similar themes and all with didactic purposes. Their backgrounds and beliefs could hardly have been more different. Mrs Hall, living in London, yet anxious to be identified as Irish and eager to improve both the image of Ireland and the behaviour of its inhabitants, honoured for her 'femininity' and purity as a writer, a convinced Protestant but a believer in religious tolerance; William Carleton, a true peasant from County Tyrone, renowned for his portraits of peasant life, a Protestant who had once been a Catholic, and a scourge of organised religion; and Charles Lever, the writer of military yarns, a doctor and a sophisticated worldly man who generally kept aloof from religious and political controversy. From their very different viewpoints they all saw the same scene and were impelled to write about it, in order to teach important lessons to those whom they knew to be most in need of them – the landlords of Ireland, and the political system which allowed them to neglect their duties and to abuse their powers.

TEN

Assessments – then and now

Mr and Mrs Hall were extremely hard workers – between them they accounted for over 400 publications – but they did not neglect the social side of life. They entertained their friends and colleagues to dinners, suppers, teas and receptions in their home, or rather homes, for they moved around London quite a lot. The Hall letters are headed from various London addresses but are all, unfortunately, undated. The only address for which we have a definite time span is that of The Rosery in Old Brompton, Kensington – 1839 to 1849 – and it was certainly here that many of the social gatherings were held. In his autobiography *The Later Years*, Henry Vizetelly, the publisher, recalls meeting Jenny Lind, the Swedish soprano, 'at a reception in that little doll's house, the Old Brompton Rosery'. The singer was so famous, and so much sought-after that a chance

to see her in private, and with a chance, too, of speaking with her, was an irresistible attraction to the school of people Mrs Hall had invited. It was one of those assemblages at which the guests commonly spend an hour or two on the landing and then retire disgusted. Getting in and out of the suite of little rooms at the Rosery was scarcely to be accomplished by any amount of struggling, and I long remained hemmed in near the lady whom everyone was so eager to see and pass their opinion upon.[1]

Thomas Moore was a lifelong friend of the Halls, 'those clever writers upon Ireland'[2] as he called them, and was a frequent visitor to The Rosery. There is an amusing description of the poet in the autobiography of W.P. Frith, the artist, who remembers seeing the poet

in the pretty little cottage in Brompton called The Rosery. It was there, on a very hot night in the height of the London season, that I saw – for the only time in my life – a lion thoroughly lionized. The lion was Tom Moore the poet; and the lionizers, consisting chiefly of ladies, clustered round the little man and nearly smothered him. Moore was so diminutive that I could scarcely see his small gasping mouth, which in its efforts to inhale the dreadful atmosphere, reminded me of a fish out of water. No wonder that he lost one of his shoes; and it was "a sight" to see him sitting, like one of Cinderella's sisters, whilst a very pretty admirer insisted on replacing the shoe on his little foot.[3]

The Halls moved from The Rosery in 1849 and the next address of theirs for which we have a date is Bannow Lodge, The Boltons, Brompton. They were there in 1861 (it was from that address that Mrs Hall edited *St James's Magazine*) and the social gatherings were still a feature of the Halls' life, for in the *Illustrated News of the World* in 1861, there is a reference to Mrs Hall's receptions at Bannow Lodge. It is in a short article accompanying an engraving of a photograph of Mrs Hall in the series 'The Drawing Room Portrait Gallery of Eminent Personages'. This reads: 'Mrs Hall occupies a high position in society, and notwithstanding the pressing demands necessarily made upon her time by the numerous literary engagements into which she has entered, manages to find frequent opportunities to "receive" and "be received" at her charming residence, Boltons, Brompton – Bannow Lodge'.[4] The photograph is that of an extremely handsome woman, heavily bejewelled, as indeed she is in the portrait of her painted in 1851 by G. de Latre and which now hangs in the National Gallery in Dublin.

In a tribute to Mrs Hall, written after her death in 1881, and reproduced in *The Life and Death of Llewellynn Jewitt, F.S.A.* by W.H. Goss (Jewitt and his wife were very close friends of the Halls), the blind writer Alice King speaks of Mrs Hall's genius for entertaining. Her drawing room was

one of those favoured rooms where a spell of ease and freedom seems always to be at work; drawing everyone that enters under its beneficient influence, and yet harmonising all into one blended whole, whatever widely differing elements may be there ... [There] met all the wit and genius which through more than fifty years, made the world laugh and weep, and sent streams of amusement and instruction flowing hither and thither in the land.[5]

Alice King lists some of the guests – Charles Lamb, Samuel Coleridge, William Wordsworth, Letitia E. Landon and Mrs Henry Wood. Mr Hall, too, in his own autobiography, gives a long list of literary celebrities who enjoyed his hospitality, but, regrettably, none of them seems to have thought it worth mentioning in their memoirs. John Forster, however, in his Life of Charles Dickens, lists the celebrities he had met in Dickens's house – Frith, Ward, Helen Faucit, Sims Reeves, Thomas Ingoldsby, Charles Knight, and adds 'and I have met at his table Mr and Mrs S.C. Hall'.[6] Samuel Carter

Hall, in his memoirs, states that he knew Dickens as a young boy when he, Hall, was a parliamentary reporter, and also 'in the early days of his celebrity'. The Halls were present at the christening of Dickens's first-born, and there is a reference in the memoirs to a children's party in the Dickens's household where the Halls met Fanny Hood, daughter of Thomas Hood.[7] No date given, but it was after the poet's death, which occurred in 1845, and, if we are to believe a story in Peter Ackroyd's life of Dickens, it would have been before 1858, when according to what Ackroyd himself calls 'malicious gossip' the novelist was snubbed by the Halls: 'Dickens and his family called upon the Samuel Carter Halls and invited them to Tavistock House, only to be told that "... that it was with Mrs Dickens they were acquainted – that if Mrs Dickens were at Tavistock House they would be happy to call, but otherwise – afraid – very sorry – but, etc. etc."'[8] Hall himself makes a passing comment on the Dickens's marital difficulties, saying that 'sympathy was largely felt for Mrs Dickens, and rightly so', but goes on to say that public comment is unseemly and that Dickens's own lack of reserve is to be deplored.[9]

Whether or not the Ackroyd story is true, it is of interest, reminiscent as it is of the Halls' earlier doubts about the propriety of calling on the Countess of Blessington when her London ménage was a source of scandal. They resolved the problem in their own way, as Mr Hall recalled in his memoirs: 'Mrs Hall never accompanied me to her evenings, although she was a frequent day-caller', adding, 'We were not of rank high enough to be indifferent to public opinion'.[10] Circumstances had changed however, and the Halls would have had sufficient self-confidence and self-esteem to make their views and their sympathies very clear, even if it involved snubbing such an important literary figure as Charles Dickens. Mr Hall, as editor of the influential *Art Union* (the title under which the *Art Journal* was published from 1849), was the arbiter of artistic taste in certain English circles. The collector Robert Vernon had granted him permission to engrave his pictures and publish them in the *Art Union* before the collection was handed over to the National Gallery of England. Hall also received a mark of royal favour when Queen Victoria and Prince Albert gave him permission to engrave 150 pictures from their private collection for publication in the *Art Union* in 1851 for a special edition on the Great Exhibition. The circulation of the periodical increased significantly but the cost of production was so high that Hall was forced to sell his share of the *Art Union* to his co-proprietors and from then on he remained a salaried employee until his retirement in December 1880. (S.C. Hall never seemed to have luck with his ventures, and never seemed to learn from experience: a similar episode took place in 1875 when he lost a large sum of money on the publication of temperance tracts which were too expensively illustrated for their selling price.) Nevertheless he was influential and well-known and could afford to make clear what his standing was on matters of morality. Hall had his detractors, notably in *Punch*, where fun was made of his enlightened efforts to raise the standard of English industrial design and craftmanship. It was supposedly Douglas Jerrold of *Punch*, who nicknamed Hall Mr Pecksniff, after Dickens's character in *Martin Chuzzlewit*. In a typical *Punch* article in 1846 the Halls' home in Brompton is referred to as the 'Pecksniffery'[11] and is described as being furnished with household items donated by grateful manufacturers in return for publicity in the *Art Union*. Presumably the charge of hypocrisy was made because of this supposed, or alleged, lack of artistic and journalistic integrity because no hint of domestic scandal ever touched either of the Halls.

Mrs Hall was also a successful figure in her own right. She contributed many articles to her husband's *Art Union*, some of which were later reprinted in book form (*Midsummer Eve*, *Baronial Halls Pilgrimages to English Shrines* and *The Book of the Thames*, etc.), but she also continued to write stories for *Chambers's Journal* and for Chambers's 'Miniature Library of Fiction'. Much of her work was for children; some of it specifically for her own 'dear little Fanny', Thomas Hood's daughter, with whom the Halls formed a lasting friendship. Mrs Hall also edited *Sharpe's Magazine* and later the *St James's Magazine*. Her earlier works, especially the Irish ones, were often reprinted, and numerous collections of her stories appeared in different forms. She remained in the public consciousness as an Irish writer, although the only new Irish novel published between 1845 (*The Whiteboy*) and 1869 (*The Fight of Faith*) was the whimsical *Midsummer Eve*,[12] 1847. It was a limp little tale about fairies in Kerry and artists in London, and marked a sad decline from her achievement in *The Whiteboy*, although some of the critics, including those in the *Observer* and the *Atlas*, reviewed it kindly. The *Athenaeum* reviewer was grateful that Mrs Hall had 'as usual' spared the public details of Irish crime and Irish ruins, and praised her for her 'fertility in inventing incident'. However, the reviewer was not impressed by the fairies who fluttered through the pages. They were not what one expected to find in Ireland, but were 'Brompton elves'.[13]

Mrs Hall obviously worked hard at her writing, and put a great deal of energy into her socialising; she also devoted herself to good causes, and at times she put her pen and her social energies at the service of philantropy. She and Mr Hall, along with Sir Philip Rose, were among the founder members of the committee that set up the Brompton Hospital for Consumption (now the Brompton Hospital) and she wrote a novella, *The Forlorn Hope*, (1845) specially for the fund. She also helped to organise a concert given by Jenny Lind in aid of the Hospital (Jenny Lind, or Mrs Otto Goldschmidt, was a near neighbour while the Halls lived in The Rosery in Brompton) and helped at numerous bazaars, which she publicised in the *Art Union*. Mrs Hall was also very interested in the sad fate of governesses who were either too ill or too old to obtain employment, and she worked diligently on behalf of the Governesses' Benevolent Institution Asylum which had been set up in 1843, and wrote a story, 'The Old Governess', especially for its funds. The Nightingale Fund was another of the Halls' projects, and the money collected was used, at the request of Miss Nightingale herself, for the training of nurses. Mr Hall gives details of these charitable activities – and many more – in his *Retrospect of a Long Life*, and the pages of the *Art Union* contain many references to the good causes supported by the Halls.

Mrs Hall was identified in the eyes of her admirers with the highest feminine virtues. She was a good wife, whose literary career in no way affected the tranquillity of the home. The fact that she and her husband worked in partnership and were co-authors of several books was a distinct advantage – it was a sign of domestic as well as literary harmony, and her earnest wish to do good through her writing was recognised and applauded. In an article about Mrs Hall in the series 'Illustrious Women of Our Time' the writer in the Philadelphia periodical *Godey's Magazine and Lady's Book* ranked her 'high among those talented women on whom rests ... the responsibility of influencing the opinions of their readers, and turning them into the right channel, or subjects of no little importance'. There was high praise for *Ireland, its Scenery, Character, etc.*: it contained instruction, amusement and information, and was notable for its truth and pathos. *Stories of the Irish Peasantry* displayed all Mrs Hall's virtues of 'faithfulness, purity and right-thinking' and her books for young people were 'a pleasing combination of fancy and instruction'.[14] In this context it must be noted that Mrs Hall had no sympathy with contemporary demands for women's rights, among married women. In fact, she was opposed to divorce and wifely independence and often said so. In his *Life of Llewellyn Jewitt* W.H. Goss quotes a letter from Anna Maria Hall – that, 'truly great woman, whose name and abilities and character were honoured in all lands' – written to a young wife 'giving a woman's view of a woman's duty':

"Never, never put faith in a woman who, having knelt at God's altar, would go free of her bond, or abate her duty to the head and heart of her existence. I tell you Mary – Mary dearest, believe me – this new seeking of womanly independence among married women is an outrage against God and nature ... it is what no Christian woman can dare to countenance. She can never remove the seal from the bond. Let her beware of signing it. If she find she cannot bend, let her never enter into the covenant; but having entered, no human law can unbind – no word of man can unloose – what God has joined. Man was created to protect and cherish – woman lovingly to serve; there is no reasoning, no arguing, 'If *you* cherish, I will serve'".[15]

It is not necessary to read Mrs Hall's correspondence in order to discover her views on wifely independence or on divorce. Her published works are not only implicit in their acceptance of the sanctity of the marriage bond; they also contain several explicit statements of her belief. In a short story 'Papa's Letter', published in *The Juvenile Budget* in 1840 she addressed women on the subject of domestic happiness. They 'would do well to remember', she said, 'that all the brilliant accomplishments ... in the world, are nothing worth, in comparison to a patient and cheerful temper, and an affection for, and perseverance in, the moral and domestic duties of life. Home ought to be the temple of a virtuous female'.[16] That female had promised at the altar to obey and although one of her heroines, Geraldine in the story *The Private Purse*, had 'thought for a moment how harsh that word "obey" sounded', she loved her fiancée so much that 'obedience would be a pleasure'.[17]

Mrs Hall, however, was honest enough not to be overly romantic. Later in the same story, when Geraldine's husband misbehaved, the author stated 'let her husband's conduct be what it would, her duties, solemnly pledged at the altar, remained the same'.[18] The same belief was held by Mrs Mansfield, one of the wives in the story *Wives and Husbands*, who knew that no matter how a husband behaved, 'the wife was bound to fulfill her part of the contract'.[19] There is no suggestion that Mr Hall ever behaved badly, or was less than honourable and loving but he *was* human, and no doubt the marriage had its share of normal irritations. It was written of him that although he 'won much love, it was not by a spirit of general concession that he won it. He was ever a man of great firmness in pursuing his own idea of what was right, and not yielding to that of the friend who differed from him'.[20] Could that trait have been in Mrs Hall's mind when she wrote

It is seldom that a woman, resolved to bear and forbear, cannot succeed in winning her husband's friendship in the end. It is wiser for her not to complain of him she has sworn to love, honour and obey... A firm and noble [heart] *will* bear it, because it is right, and perhaps, after years of firm endurance, be rewarded by the friendship it has so richly deserved – the friendship of *him* in whom a young heart trusted.[21]

Romance gives way to friendship, after years of endurance, but never is there any suggestion that a husband and wife should lead separate lives, or that they should be parted, except by death, and there is, nowhere in Mrs Hall's writings any demand for wifely independence, or for women's rights. In *The Private Purse* Geraldine, 'happily for herself, had never thought of discussing the rights of women apart from the rights of men'[22] and that was all Mrs Hall wrote on the subject. It is strange so, that the entry for Mrs Hall in *The Bloomsbury Dictionary of Women Writers* describes her as a campaigner for women's rights,[23] but possibly this is no more than a repetition of the entry in the *Dictionary of National Biography* which makes the same statement.[24]

That false assertion is also made by Michael Scott, editor of the condensed version of *Halls' Ireland* which appeared in 1984. In his Introduction he states that Mrs Hall 'was an early champion of women's rights and wrote many tracts for the cause'.[25] Needless to say, no such tracts are to be found. A possible explanation of the original biographical error, may have been a misreading of a passage in Goss's book on Llewellyn Jewitt which said

The wives of Samuel Carter Hall and Llewellyn Jewitt were both staunch believers in the maintenance of women's rights, but never had any occasion to make any loud demonstration on the subject. They received and exercised these rights as a mere matter of ordinary course. Had the rights of either wife to be loved, cherished, honoured and protected by man – and that man her husband – to be challenged she would have been astounded.[26]

The women clearly believed that they had rights as wives, rights which they fiercely upheld, but these rights did not include independence, nor were they rights which should be extended to unmarried women, nor to women who had broken their marriage vows.

As Mrs Hall's writing career continued until shortly before her death the question to be asked is why did she write no more works of literary significance? The answer is almost certainly because she never again wrote at any length about Ireland. Her earlier books came to life only when she was writing about 'her' native country and 'her' countrymen and countrywomen and once she ceased to write on this subject her talent deserted her. One can only speculate on *why* she gave up writing about Ireland, but two reasons come to mind. In the first place, Mrs Hall had been away from Ireland too long and had already exhausted her stock of Irish memories. She visited the country often, it is true, but she could never again be part of it, and although she was a sharp and sympathetic observer while she was touring or staying with friends, it was not the same as living among Irish people. The other reason, of course, is the enormous change that had come about in Irish life and in the English perception of Ireland because of the great famine of 1845–1849 and its aftermath – large-scale emigration. English public opinion had been horrified by the stories that came from famine-stricken Ireland and there had been an upsurge of public sympathy and a flurry of charitable activity among all classes during those terrible years, but it was not the right atmosphere for the enjoyment of bright little Irish tales, or, indeed, any tales about Ireland.

As early as 1847 there were indications that the public was growing weary of Irish subjects. In its review of Anthony Trollope's *The Macdermots of Ballycloran* the *Athenaeum* complained 'Banim, Griffin and Carleton have laid before us so many tragedies of dull domestic misery, or of those sharper agonies that destroy reason and life that "Memory shudders at the dreary tale"; and an Irish Novel has become to us something like the haunted chest in the corner of Merchant Abudah's apartment, which even when closed he knew to contain a shape of Terror and a voice of Woe'.[27] A masterly novel about the famine might, or might not, have caught public and critical interest, but Mrs Hall was not the person to write it. Neither, as it transpired, was William Carleton. His famine novel *The Black Prophet* (1847) although it contained many powerful and moving passages was poorly constructed, and was confused and incoherent in places, while his later work, *The Squanders of Castle Squander* (1852) was so burdened with facts, figures, and overt didacticism that his chilling descriptions of death and disease were almost obscured. Mrs Hall was aware of her limitations and in a full-page article in the *Art Union* in 1847 she says simply that she can 'devise no fiction equal to the fact – the fearful realities of death and starvation' that are conveyed to her by every post. In this article, 'The Cry From Ireland', Mrs Hall lays blame for the calamity where she has laid it before: on the English government for mismanagement, misrule and apathy, on absentee landlords, the absence of capital for industry, and on profiteering Irish farmers who send food out of the country while their fellow men and

women are dying of hunger. She is appealing for money to help
relieve distress in Ireland and explains why she is using the pages of
a journal devoted to art for this purpose. The readers of the
periodical know of her Irish connections, and, she goes on: 'I would
fain believe that in drawing, as I have often done, portraits of the
Irish character to interest and amuse them, I have excited sympathy
and affection for the Irish people; and I venture to hope that some
will make me the almoner of their charity'. She is looking for small
gifts, those that are too small to be called public subscriptions, and
she promises that she can 'bestow them where they would be
properly applied', adding that 'it would now be difficult to apply
charity improperly in Ireland'. It is a fine article, full of indignation,
regretting that because of past mistakes, all that can now be done for
the suffering people of Ireland is an alleviation of their distress
through charity. She emphasises the fact that this cry for help comes
not from 'the wild Indian' or the 'dull Hottentot' but from those
who are England's 'nearest neighbours ... closest fellow-subjects ...
most natural allies ... a people full of the elements of good; with the
kindliest sympathies, the warmest affections, the most enduring
fortitude – forbearing and honest to the death'. There is optimism in
the article – an optimism which is very much Mrs Hall's own.
Things must change, she believes: 'After what has come to pass, it is
impossible the same system of things must be suffered to continue,
property must change hands; lands must no longer be permitted to
want hands, nor hands lands'.[28] Nothing did change, or not in Mrs
Hall's lifetime, and her belief, expressed in *The Whiteboy* and re-
stated here, that England and Ireland would draw closer together
was not justified. The charitable aid sent by English organisations
and individuals was not enough to avert disaster nor did it wipe out
Irish resentment at what was seen as English governmental
callousness. Neither did these generous efforts 'set the brand of
shame on the forehead of the agitator' who would seek to separate
the two countries. In a wasted land the seeds of separatism were
below the surface, but they were dormant, not dead, and Mrs Hall
lived to see them spring up first, very briefly, in 1848 in the Young
Ireland Rebellion, only to wither and then flourish years later as
Fenianism.

Mrs Hall, however, continued to be optimistic about the future
relationship between England and Ireland, and in a new edition of
The Whiteboy, issued in 1855 – she introduced the novel with brave
and hopeful words: 'England has learned to treat Ireland, not alone
with justice, but with sympathy: various circumstances have
combined to render "the bit of land" no longer an object of

perpetual death-struggle: the landlord has grown more considerate, and the labourer better contented – because better treated and better recompensed'.[29] The great famine was, no doubt, one of the 'various circumstances' which had altered the situation of landlord and tenant in Ireland, but Mrs Hall did not advert to it, or to its consequences. Perhaps the horror was too great for her to bear, for she was a sensitive woman who loved Ireland and the Irish in her own way, and that she could not bring herself to record the facts, and as she said herself, fiction could not do justice to the calamity. Instead, like many another commentator of the period she took refuge in the belief that out of such evil must come good. The passage on the Poor Law System in Ireland which appeared in a later edition of *Halls' Ireland* and which was quoted earlier in Chapter VII of this study throws some light on the Halls' attitude towards the famine of 1845–1849. It combines their unshakeable optimism with words of reproach and a reluctance to dwell upon the horrors of the grim years. Those details were hardly suitable for a book designed to attract visitors to Ireland so the Halls, in effect, avoided the issue.

In 1869 Mrs Hall produced a novel, *The Fight of Faith* which she described in the Dedication as her last work of fiction. 'With these volumes', she said, 'I bid the public a grateful farewell', and then went on to make clear her motives for writing the book. It was an attack on the proposed disestablishment of the Church of Ireland, which, Mrs Hall, like many another, saw not only as a betrayal of Irish Protestants, but the start of a process of Romanisation that could damage the social, political and religious life of the two islands. 'It will be cause for thankfulness and happiness' she wrote 'if I can, in any degree, arrest the progress of those who are seeking to negative the blessings brought to the Kingdoms by the Reformation, and by that Protestant Ascendancy to which we owe so much of our liberty and so many of our rights.'[30] This is a reiteration of what the Halls had already said when describing the Williamite wars there, in their guide to Ireland, and can be accepted as an expression of honest belief, but the tone of the book in general is unpleasantly bigoted. The plot concerns the sufferings of a Huguenot girl who flees the persecution in France, lives for a time in England, and then travels to Ireland, where the narrative concludes with a description of the Battle of the Boyne. The language used is lurid: the success of Roman Catholicism in England would involve 'the lives of men, women and children offered to a huge idol, that seated knee-deep in blood, smiled at each new sacrifice'. The 'pollution of Anti-Christ' would spread over the country as it lay 'in the grasp of Rome' and The Scarlet Woman would walk the land, breathing forth 'the persecuting spirit of Popery'. In Ireland, King William's forces would have to 'rescue Protestants from the foul grasp of Papists' and deliver them from their tortures under 'the harrow of Rome'. An extreme, and almost comical, comparison is made between the Catholic Irish maid, Nelly and her Huguenot mistress, Pauline. Nelly is 'a stout, strong-limbed Irish girl, her large scarlet mouth half-open' and her earthly beauty is crude beside Pauline's spiritual loveliness. There is a clear echo here of Mrs Hall's earlier descriptions of coarsely attractive Irish peasant girls and their delicately beautiful young mistresses, but now it is not only upper-class background that is seen as a source of physical refinement – the Protestant religion also confers that advantage.

Mrs Hall's earlier religious tolerance has quite vanished by now, and she has settled for the clichés of bigotry. Not surprisingly, the book is almost impossible to read, loaded as it is with rancour. It is poorly-plotted and badly-written, and has none of the freshness and humour that characterised some of her earlier writings. The central belief, that the Church of England (and its sister, the Church of Ireland) is the true and reformed church of Christ is nothing new in Mrs Hall's philosophy – she had always made it clear what her religious views were – but the personal attacks on adherents to other faiths are new. In one of the few legible letters remaining to us, and written in 1853 to her friend Francis Bennoch, Mrs Hall had stated categorically, when speaking of the possible employment by a friend of a governess who was a Roman Catholic that she 'would not give as much influence to any Roman Catholic in a Protestant family', and added, 'I like people to be firm and faithful to whatever creed they profess' pointing out that there was a danger that such a person might, from the best motives, seek to proselytise. She concluded, 'There are some RC's whom I love, but I would give them no domestic influence where there are young persons'.[31] Given Mrs Hall's beliefs this was a reasonable point of view, and makes the subsequent slide into bigotry particularly sad. An apt comment on *The Fight of Faith* was made in a review which stated 'The Battle of the Boyne is depicted with vigour and spirit but we can not imagine it will do much good to revive these memories', adding that as Mrs Hall was so opposed to disestablishment and disendowment she had felt it her duty to write such a polemic.[32] In the *Athenaeum*, however, a review of *The Fight of Faith* contains the assertion that the novel was written some years previously and 'was not called forth by any events or considerations of the present day'. The reviewer complained about the confused narrative, because 'Mrs Hall has sacrificed her story to give fuller details of the Fight of Faith in Ireland between King James and King William', but nonetheless, this book was, 'in many respects her best'.[33]

The *Fight of Faith* was indeed Mrs Hall's last novel but she continued to work as a journalist until shortly before her death in 1881 and two later works have some Irish interest. A pamphlet *God Save the Green*, subtitled 'A Few Words to the Irish People' appeared about 1871 in a series 'Illustrated Readings', and was strongly anti-Fenian. Mrs Hall pointed out that armed risings always hindered progress in a country, adding with her usual optimism, that conditions in Ireland were much better than they ever had been and that the future looked bright. The Chief Speaker was a Roman Catholic, was not this proof of a new and happy era?[34] This was Mrs Hall's last comment on Irish affairs, but in a booklet by Mr Hall 'A Memory of Thomas Moore', which was produced in order to raise funds for a memorial window in Bromham where the poet was buried, Mrs Hall recalled her meetings with Moore and his wife. In a review of Mr Hall's 'little work' the *Irish Builder* praised Mrs Hall not only for her chapter in the booklet but for her work in the past. 'She contributes,' said the reviewer, 'to the interest and perfection of the "Memory", and it has been a joy to more than one generation of our people that Mrs Hall has lived to see the pleasing fruit of her labours in many literary walks receiving a practical embodiment – for has she not preached for more than an ordinary lifetime the lessons of self-help, self-exertion, thrift, economy, punctuality in her numerous racy tales and sketches of Irish life and character?'[35] The 'more than ordinary' lifetime came to an end on 30 January, 1881, when Mrs Hall died in her home, Devon Lodge, East Molesey, Surrey. She was buried in the churchyard at nearby Addlestone, where an ivy, brought years earlier by the couple from Killarney, climbed vigorously over the church tower. Tributes to the late author appeared in several publications and were generally respectful in tone. *The Times* referred to her as 'a well-known authoress' who had written Irish sketches, numerous books for children and worked hard in the cause of temperance.[36] The obituary wrongly gave her birth-date as 1807, an error which was repeated in the *Irish Times* in its notice. The *Irish Times*, referring to Mrs Hall's *Sketches of Irish Character* noted that 'Irish subjects at the time possessed all the charm of novelty' and added that she 'did some very good work in collaboration with her husband'.[37] The *Wexford Independent* in a long and fulsome article wrongly stated that she had been only seventeen at the time of her marriage to a 'spouse who introduced her to a literary career'. The obituarist, who claimed to be a close friend of the

Halls, praised the 'heartfelt interest [Mrs Hall] ever manifested in the happiness and welfare of her native land, and in the moral, social and religious training of its children'.[38] The writer of the obituary in the *Irish Builder* concurred, saying that this 'very voluminous writer' had written stories that 'inculcated virtues of industry, thrift, punctuality, sobriety and never putting off until tomorrow what could be accomplished today'.[39] The *Athenaeum* asserted that 'the best parts of her novels are those that deal with Irish scenes and characters' adding that 'she may be considered one of the most successful of Miss Edgeworth's followers'.[40]

Samuel Carter Hall lived for another eight years, mourning his wife, but consoled by the contacts he made with her through spiritualism. After his death, in March 1889, several obituaries were published and one, in the *Irish Builder* summed up his literary work neatly, saying that it was 'of a useful, more than of a brilliant nature ... 'His wife', added the writer, was 'an author perhaps even more distinguished than, and quite as laborious as himself'.[41]

Later assessments of Mrs Hall's work have been varied. Horatio Krans in *Irish Life in Irish Fiction*, published in 1903, was dismissive, seeing 'her work in Irish fiction ... but one outlet for her zeal for moral and social edification', and quite 'without Miss Edgeworth's wit'.[42] Stephen J. Brown, S.J., on the other hand, (*Ireland in Fiction*, 1918) admired her use of dialogue which was 'very well done, full of humour and flavour', and considered that as 'a graphic delineator of Irish life and character few writers ... dealt with the subject so delightfully and truly as Mrs Hall'.[43] (This was certainly not the view of an earlier commentator, a Mr J.C. Twomey who wrote in 1850 in a paper for the Kilkenny Archaeological Society that she was 'a caricaturist of the Irish peasantry'. He based this statement on interviews with people in Bannow who remembered her kindly as a person but who resented what she had written about local characters.[44]) Mrs Hall's work is not mentioned in Thomas Flanagan's *The Irish Novelists* published in 1958 (although he quotes from 'the owlish S.C. Hall', whom he describes as having 'for his sins, spent some time in the country' – a rather inadequate description of Hall's relationship with Ireland) and there is no reference to *The Whiteboy* in James M. Cahalan's study of Irish historical novels, *Great Hatred, Little Room*, published in 1983. James Newcomer, however, in an essay in 'Etudes Irlandaises' in December 1983 describes *The Whiteboy* as a 'large, generous and well-shaped novel'.[45] Mrs Hall, he concludes, shows herself as 'perceptive, accurate and sympathetic' in her analysis of Anglo-Irish problems of the period and he sees the novel as an important work with relevance to present-day Irish conflicts. Barry Sloan in his *Pioneers of Anglo-Irish Fiction 1800-1850* published in 1986, gives a detailed analysis of the novel, and while he concedes that it is 'seriously flawed by absurdities in the plot and the author's compulsive moralizing' he nonetheless considers it of some literary value as well as having authorial insights that were unusual at the time.[46] In an earlier essay by Sloan, 'Mrs Hall's Ireland', published in *Eire/Ireland* in 1984, he had noted her 'real originality' in taking peasant characters as subjects for her stories, but found that it paled beside Carleton's treatment of those subjects. 'She might', said Sloan, 'have appeared a more important pioneer had it not been for the publication of Carleton's stories at almost the same time.'[47] Gregory Schirmer in his contribution to *The Irish Short Story*, published in 1984, believed that Mrs Hall's ear for dialect was good, although not as good as Carleton's, and he praised her 'ability to present accurately and honestly many aspects of her province'. Her sketches, he said, 'although less innovative than Maria Edgeworth's *Castle Rackrent*, constitute a ... significant addition to the Anglo-Irish tradition'.[48]

The last collection of Mrs Hall's early Irish sketches was published in Edinburgh in 1913, but in 1979 the Garland Publishing Company (New York and London) produced facsimile editions of *Sketches of Irish Character*, First Series, *Lights and Shadows of Irish Life*, *Stories of the Irish Peasantry*, and *The Whiteboy*. In his introduction Robert Lee Wolff noted Mrs Hall's limited understanding of the problems of Ireland, but he praised 'the generosity of her thinking'. She was, he stated, an 'unexpectedly enlightened and eloquent member of her own generation'.[49] It is this cast of mind that makes Mrs Hall such an interesting writer. She was a condescending colonial in her attitude towards the Irish peasantry, it is true, and was firmly fixed in the class consciousness of her times, but she was not guilty of the contemporary sin of religious bigotry. Peasants may have revolted her by their dirt and their drunkenness, and she was intolerant of their fecklessness, laziness and imprudence, but she did not believe, unlike many others who wrote at the time, that these faults were a direct result of their Roman Catholicism. It was only at the end of her life that she succumbed to sectarian hysteria and *The Fight of Faith* is as much a response to the threat of Romanism in England as it is to the question of Irish disestablishment. Wholesale conversion to Protestantism by Irish peasants would not have cured them of their faults – they could prosper within the confines of their own religion, provided they took advice and guidance from their betters. Mrs Hall genuinely believed in the great potential of the Irish people, of all classes and creeds (although being a Protestant was, for many reasons, a great advantage), and for that alone she is to be respected. It is too much to expect a writer who was not a genius to break free from the confines of class, religion and background, and Mrs Hall was no genius, but a sympathetic observer who recorded, in a fresh and lively fashion, what she saw and heard in Ireland and commented upon it. Her comments reveal her limitations, but she was less limited than others of her time. The didacticism is irritating to the modern reader, but was acceptable to her contemporaries – was, indeed, expected – and I believe that that seriousness of purpose lifts Mrs Hall's work from what otherwise would have been a series of sketches like those of Miss Mitford; pleasant, enjoyable, but not memorable, except in a most superficial sense, to a more important place in the history of Anglo-Irish literature. The didacticism, along with her 'Irishness' was one of the reasons Mrs Hall was such a popular author in her time. Another reason for her success is a simple one – her writings were easy to read, literally so, because although she did use some Irish dialect in recorded speech it rarely became the 'jargon and uncouth orthography' that had been criticised earlier by a reviewer of William Carleton's *Traits and Stories*. The critic (in *Tait's Edinburgh Magazine*) complained about 'the tiresome, parrot-like repetition of some bald Irish word or phrase regularly explained at the bottom of the page – till the pages look more like sounding dictionaries than compositions intended to be made descriptive or racy by the use of frequent phrases and picturesque native words illustrating the genius of a people through their language. We humbly submit that there is neither wit, humour or feeling in lots of superfluous h's , in g's lopped away or double ee's broadened into a's'.[50] The sense of place in Mrs Hall's Irish sketches is very strong, particularly so when she is writing about her native Bannow, where the sea is a constant presence, and wide horizons and high skies give a feeling of spaciousness and freedom. Her characters may in some cases be very crudely sketched and may appear insubstantial, but their physical world is a very real one, with details of their houses, their food, their furnishings and their friends and neighbours. If it were not for the fact that Mrs Hall included so many violent features in her work one would say she worked in water colours or in pastel, but as it is, the nearest visual equivalent is enamel. There is a vividness and sheen to what she does, and although it is on a small scale she succeeds in composing her picture so that it catches the eye and lingers in the memory. Her prose is effortless, flowing easily along and with the power of summoning up images in a few words. Occasionally she slips into banality and pathos, and sometimes quivers on the edge of melodrama but it is noticeable that when she is absorbed in her efforts to teach the world about Ireland her prose is direct and clear-cut.

It is no surprise so, that Mrs Hall should have been a success as an 'Irish' author, and that she should have been accepted as an interpreter of Irish ways. Sadly, it can be argued that her very success did 'her' countrymen and women a disservice. The vividness of her Irish sketches and stories made them memorable, and her images of Ireland were those that many people in England would have retained throughout their lives. Some of these images (reinforced by the illustrations by Maclise, that appeared in later editions of *Sketches of Irish Character*) were of ragged, raw-boned hulking men who either lounged lazily on gates, or sat smoking pipes in filthy hovels while their slatternly wives thrust potatoes into the mouths of half-naked children and plump pigs. To counter-balance these images there were those of respectably-clad men and women, paying discreet and grateful homage to the landlord or his good lady who had lifted them up from the mire. The subliminal message was clear, Irish people *can* be saved, but they need help from others – they cannot do it on their own. Cheap literature is in many ways astonishingly potent, perhaps because of the ease with which complex fictional problems are solved by simple means and it is perfectly possible that men of power and influence in England may have been affected in their young days by Mrs Hall's rhetoric and were subconsciously swayed by it when dealing with the 'Irish problem'.

Mrs Hall's books are not now read except by students of that period of Anglo-Irish literature (and those who are studying minor Victorian women writers in England) but the work she published in conjunction with her husband, Samuel Carter Hall, the three-volume guide *Halls' Ireland* is still familiar to many. The condensed two-volume edition, edited by Michael Scott, and published in 1984 by Sphere Books is recommended to tourists visiting Ireland, not only for its 'idiosyncratic, highly personal account of Ireland in the last century' but because it is possible to retrace the route the Halls took and to see many of the sights they saw, some of which remain the same. Mrs Hall, so, if she is remembered at all, is linked in memory with her husband, which is ironic, considering that she was the better writer of the two, but when we recall her frequent assertions that the career of a wife was unimportant compared to that of a husband, then, by her lights, she would have seen this as being entirely proper.

BIBLIOGRAPHY

(A) WORKS BY MRS S.C. HALL
Hall, Anna Maria,
The Boy's Birthday Book, 1859.
The Adventures ... of B. Dorking, edited by Mrs S.C. Hall, 1859.
Stories of the Flowers, with an introduction by Mrs S.C. Hall, 1877.
National Temperance League. *Woman's Work in the Temperance Reformation,* with an introduction by Mrs S.C. Hall, 1869.
The Gift of Friendship, with contributions by Mrs S.C. Hall, 1877.
Hall, S.C. and Mrs A.M.,
The Book of South Wales, etc., 1861.
The Book of the Thames, 1859.
A Companion to Killarney, etc., 1878.
Hand Books for Ireland, etc. 1853.
Ireland: its Scenery, etc., 1841, etc.
Tenby, etc., 1869.
A Week at Killarney, 1843.
Hall, Anna Maria,
Finden's Tableaux [for 1837], edited by Mrs S.C. Hall, 1837.
The *Juvenile Forget Me Not,* edited by Mrs S.C. Hall, 1829.
The *St James's Magazine,* conducted by Mrs S.C. Hall, 1861.
Sharpe's London Magazine, etc., conducted by Mrs S.C. Hall, 1845.
The Princess Alexandra Gift Book, [being contributions by Mrs A.M.H., 1868.
Alice Stanley and Other Stories, Edinburgh [printed], 1869.
[Another edition] London, 1873.
All is not Gold that Glitters. A Tale. Miniature Library of Fiction. vol. 2. 1858.
Annie Leslie and Other Stories, London, 1877.
The Book of Royalty. Characteristics of British Palaces. The drawings by W. Perring & J. Brown, London, 1839.
Boons and Blessings. Stories and sketches to illustrate the advantages of temperance, illustrated, etc., London, 1875.
The Buccaneer. A Tale. Standard Novels, No. 79, 1831.
The Cabman's Cat. (Extracted from the *British Workman*) London, 1865. No. 1 of 'Kindness to Animals' Series.

Can Wrong be Right? A Tale. 2 vols. London, 1862.
Chertsey and its Neighbourhood. pp. iv, 64, London, 1853.
Chronicles of Cosy Nook. A book for the young, London, 1875.
Cleverness. A tale. Miniature Library of Fiction, vol. 6, 1858.
Daddy Dacre's School. A story for the young, London, 1859, [1858].
A Descriptive Sketch ... of the Engraving of the Village Pastor from a Picture by W.P. Frith. pp. 8, Lloyd Brothers & Co., London [1860?]
Digging a Grave with a Wine Glass. [With illustrations], London, [1871].
The Dispensation. The Tale Book, 1859.
The Drunkard's Bible. (The Drunkard's Wife. The Drunkard is Our Brother) London, 1854.
Fanny's Fancies. The Magnet Stories, etc. [1860, etc.).
The Fight of Faith. A Story. 2 vols, London, 1869.
The Forlorn Hope. A Story of Old Chelsea, p. 28 [1846].
'God save the Green!' A few words to the Irish People. Illustrated Readings [1866, etc.].
The Governess; a Tale. Miniature Library of Fiction, vol. 1, 1858.
Grandmamma's Pockets. Chambers's Library for Young People, 1848, etc.
The Groves of Blarney, a Drama, in Three Acts [and in prose], etc., London, [1836?].
The Hartopp Jubilee; or, Profit from Play, Darton & Clark, London [1840?]
The Irish Agent. Philip Garraty: or, 'We'll see about it.' Characteristic Sketches of Ireland and the Irish. By Carleton, Lover, and Mrs Hall, etc. 1845.
The Juvenile Budget: or, Stories for Little Readers ... With illustrations by H.K. Browne, pp. x, 304. Chapman & Hall; A.K. Newman & Co., London, 1840.
The Juvenile Forget-Me-Not. Edited by Mrs Samuel Carter Hall, London, 1862.
The Last in the Lease. Seven Tales by Seven Authors, etc., 1849.
Lights and Shadows of Irish Life, 3 vols, London, Colburn, 1838.
Little Chatterbox: a Tale ... illustrated by J. Absolon. p. 19. W.S. Orr & Co., London, 1844.
The Lucky Penny and Other Tales, London, 1857.
Mabel's Curse. A Musical Drama, in Two Acts [and in prose and verse]. Duncombe's Edition [of the British Theatre], vol. 28.
Mamma Milly. Magnet Stories, etc. No. 3, [1860].
Marian; or, A Young Maid's Fortunes, 3 vols, London, 1840.
[Another edition.] The Parlour Library, vol. 9, 1847, etc.
Midsummer Eve: A Fairy Tale of Love, Longman, London, 1848.
Nelly Nowlan and Other Stories, London, 1865.
Number One: a Tale ... Illustrated by J. Absolon, p. 24. W.S. Orr & Co., London, 1844.

Number One – Perseverance. A Book of Stories for Young People, etc. 1848.
The Outlaw, an Historical Romance. Standard Novels, etc. No. 105, 1831, etc.
Pilgrimages to English Shrines ... With notes and illustrations by F.W. Fairholt, London, 1850.
Second series, London, 1853.
The Playfellow and Other Stories. [With illustrations] London, Edinburgh [printed], 1866.
Popular Tales and Sketches, Lambert, London, 1856. One of a Series entitled: 'Amusing Library for Young and Old.'
The Private Purse, and Tattle. Tales. Miniature Library of Fiction vol. 3, 1858.
The Rift in the Rock. A tale. The Rainbow Stories, No. 2, [1871, etc.].
Ronald's Reason; or, the Little Cripple, London, [1865]. One of the 'Children's Friend' series.
St. Pierre, the Refugee: a Burletta in Two Acts [and in prose], London, 1837.
Sketches of Irish Character, 2 vols, Westly and Davis, London, 1829.
Second series, London, 1831.
Illustrated edition, London, 1855, [1854].
Stories of Governesses, London [1852].
Stories of the Irish Peasantry. Chambers's Instructive and Entertaining Library, 1848, etc.
The Swan's Egg [A Tale], pp. 157, 1851. Chambers's Library for Young People, 1848.
Sketches of Irish Life and Character ...With Sixteen Reproductions from the Paintings of Erskine Nicol, p. 323, T.N. Foulis, Edinburgh and London, 1909.
Tales of Woman's Trials, London, 1835.
[Another edition] London, 1847.
There is No Hurry, and, Deeds not Words. Tales. Miniature Library of Fiction, vol. 4, 1858.
Turns of Fortune, by Mrs Samuel Carter Hall [A tale] Miniature Library of Fiction. vol. 5, 1858.
The Two Friends; a [Temperance] sketch, London [1856].
Uncle Horace, ... by/ ... [A.M.H.] 1837.
Uncle Sam's Money Box. Chambers's Library for Young People, 1848, etc.
The Unjust Judge. Miniature Library of Fiction, vol. 13. 1858.
The Village Garland. Tales and Sketches, London, 1863 [1862].
The Way of the World, and Other Stories, London, Edinburgh [printed], 1866.
The Whisperer. Chambers's Library for Young People, 1848, etc.
The Whiteboy; A Story of Ireland in 1822. 2 vols, Chapman & Hall's series of original works, etc. 1845.
[Another edition] London, 1855. One of the series entitled Select Library of Fiction.
New edition, p. 316, Ward, Lock & Co., London [1844].

Author's copyright edition. p. 159, G. Routledge & Sons, London, 1887.
William and His Teacher. The Golden Casket, etc. [1861].
Wives and Husbands; a Tale. Miniature Library of Fiction, vol. 9, 1858.
A Woman's Story. 3 vols, London, 1857.
The Worn Thimble; a Story of Woman's Duty and Woman's Influence, London, 1853.
and Foster (Mrs Jonathan). *Stories and Studies from Chronicles and History of England,* 2 vols, London, 1847.
Stories and Studies of English History. A new edition, with additions, London [1859].
Ninth edition, enlarged and brought down to the present time. Edinburgh [1866].

(B) HALL LETTERS
in National Library of Ireland and Iowa State University

(C) GENERAL BIBLIOGRAPHY
Ackroyd, Peter,
   Dickens, Sinclair – Stevenson, London, 1990.
Adburgham, Alison,
   Silver Fork Society: Fashionable Life and Literature 1814–1840, Constable, London, 1983.
Akenson, Donald H.,
   The Church of Ireland: Ecclesiastical Reform and Revelation, 1800–1885, Yale University Press, New Haven and London, 1971.
Andreas, Alexander,
   The History of British Journalism from the Foundation of the Newspaper Press in England to the Repeal of the Stamp Act in 1855, 2 vols, Bentley, London, 1859.
Andrews, Malcolm,
   The Search for the Picturesque: Landscape, Aesthetics and Tourism in Britain 1765–1800, Scolar Press, Aldershot, 1989.
Astin, Marjorie,
   Mary Russell Mitford: Her Circle and Her Books, Douglas, London, 1930.
Baker, Ernest A.,
   The History of the English Novel, vol. VII: *The Age of Dickens and Thackeray,* H.F. & G. Weather by Ltd., London, 1936.
Banim, John and Michael ('The O'Hara Family'),
   Father Connell, 3 vols, reprint of 1842 edition, Garland Publishing Inc., New York and London, 1978.
   The Anglo-Irish of the Nineteenth Century, 3 vols, reprint of 1828 edition, Garland Publishing Inc., *The Bit O'Writin',* 3 vols, reprint of 1828 edition, Garland Publishing Inc., New York and London, 1978.

Bibliography

The Boyne Water, 3 vols, reprint of 1826 edition, Garland Publishing Inc., New York and London, 1978.

The Croppy 3 vols, reprint of 1828 edition, Garland Publishing Inc., New York and London, 1978.

The Denounced, 3 vols, reprint of 1830 edition, Garland Publishing Inc., New York and London, 1978.

Barrington, Sir Jonah,
Historic Memoirs of Ireland, 2 vols, 2nd edition, Colburn, London, 1835.

Bates, William,
The Maclise Portrait Gallery of Illustrious Literary Characters, Chatto & Windus, London, 1883.

Beames, Michael,
Peasants and Power: The Whiteboy Movements and Their Control in Pre-Famine Ireland, The Harvester Press, Brighton and St Martin's Press, New York, 1983.

Beaumont, Gustave de,
Ireland: Social, Political and Religious, 2 vols, ed. Taylor, W.C., Bentley, London, 1839.

Beckett, J.C.,
The Anglo-Irish Tradition, Faber & Faber, London, 1976.
The Making of Modern Ireland 1603–1928, Faber Paperbacks, London, 1985.
Bloomsbury Dictionary of Women Writers, Bloomsbury Publishing Ltd., London, 1992.

Bowen, Desmond,
The Protestant Crusade in Ireland 1800–1870: A Study of Protestant – Catholic Relations between the Union and Disestablishment, Gill & Macmillan, Dublin, 1978.

Boyd, Ernest A.,
Ireland's Literary Renaissance, 2nd edition, Grant Richards Ltd., London, 1923.

Brittaine, George,
Irish Priests and English Landlords, Richard M. Tims, Dublin, 1830.

Brown, Malcolm,
The Politics of Irish Literature from Thomas Davis to W.B. Yeats, Allen & Unwin, London, 1972.

Brown, Stephen J.,
Ireland in Fiction: A Guide to Irish Novels, Tales, Romances and Folklore, reprint of 2nd edition, 1919, Barnes & Noble, New York, 1969.

Buckley, Mary,
'Attitudes to Nationality in Four Nineteenth-century Novelists' in *Journal of the Cork Historical and Archaeological Society*, Vol. LXXXVIII, Part I, No. 227, Jan–Dec. 1973, pp. 27–34 and in Vol. LXXXVIII, Part II, No. 228, Jul.–Dec. 1973, pp. 109–116, and in Vol. LXXIX, Part II, No. 230, Jul–Dec.

1974, pp. 129–136, and in Vol. LXXX, Part II, No. 232, Jul.–Dec. 1975, pp. 91–94.

Burke, Edmund,
A Philosophical Enquiry into the Origin of Our Ideas of Sublime and the Beautiful, ed. Boulton, J.T., Routledge & Kegan Paul, London, 1858.

Butler, Marilyn,
Maria Edgeworth: A Literary Biography, Clarendon Press, Oxford, 1972.

Carleton, William,
Art Maguire: Or The Broken Pledge, Duffy, Dublin, 1845.
Parra Sastha; Or The History of Paddy-Go-Easy and His Wife Nancy, Duffy, Dublin, 1845.
The Black Prophet: A Tale of Irish Famine, Simms & McIntyre, Belfast, 1847.
The Squanders of Castle Squander, Illustrated London Library, London, 1852.
Traits and Stories of the Irish Peasantry, First Series, 2 vols, Dublin, Curry, 1830. Seventh edition, 2 vols, Tegg, London, 1867.
Valentine M'Clutchy; The Irish Agent, Or The Chronicles of Castle Cumber, Duffy, Dublin, 1845.

Carpenter, Andrew (ed.),
Place, Personality and the Irish Writer, Colin Smythe, Gerrards Cross, 1977.

Casey, Daniel J. and Rhodes, Robert E. (eds),
Views of the Irish Peasantry, 1800–1916, Archon Books, Hamden (Conn.), 1977.

Chambers, William,
Story of a Long and Busy Life, W. & R. Chambers, Edinburgh and London, 1882.

Clark, Samuel,
Social Origins of the Irish Land War, Prince ton University Press, Princeton, 1979.

Colby, Robert A.,
Fiction with a Purpose: Major and Minor Nineteenth-Century Novels, Indiana University Press, Bloomington and London, 1967.

Coleridge, Samuel Taylor,
Miscellaneous Criticism, 2 vols, ed. Raysor, T.M., Constable, London, 1936.

Connell, K.H.,
The Population of Ireland 1750–1845, Clarendon Press, Oxford 1950.

Connolly, Seán J.,
Priests and People in Pre-Famine Ireland, 1780–1845, Gill & Macmillan, Dublin, 1982.

Craig, Maurice,
Dublin 1600–1860, Allen Figgis, Dublin, 1960.

Croker, Thomas Crofton,
Fairy Legends and Traditions of the South of Ireland, 3 vols, John Murray, London, 1825–1828.

Researches in The South of Ireland Illustrative of the Scenery, Architectural Remains, And The Manners and Superstitions of the Peasantry: With An Appendix Containing A Private Narrative of the Rebellion of 1798, John Murray, London, 1824.

Cronin, John,
  Gerald Griffin: A Critical Biography, Cambridge University Press, Cambridge, 1978.
  The Nineteenth Century Anglo-Irish Novel, Vol. I, Appletree, Belfast, 1980.

Crotty, Raymond D.,
  Irish Agricultural Production: Its Volume and Structure, Cork University Press, 1966.

Crowe, Eyre Evans,
  Today in Ireland, 3 vols, reprint of 1829 edition, Garland Publishing Inc., New York and London, 1979.
  Yesterday in Ireland, 3 vols, reprint of 1829 edition, Garland Publishing Inc., New York and London, 1979.

Curtis, L. Perry, Jnr.
  Anglo Saxons and Celts: A Study of Anti-Irish Prejudice in Victorian England, Conference on British Studies, University of Bridgeport, Connecticut, 1968.
  Apes and Angels: The Irishman in Victorian Caricature, David & Charles, Newton Abbot, 1971.

Dalziel, Margaret,
  Popular Fiction One Hundred Years Ago: An Unexplored Tract of Literary History, Cohen & West, London, 1957.

Donnelly, James S. Jnr.,
  Landlord and Tenant in Nineteenth-Century Ireland, Gill & Macmillan, Dublin, 1973.

Downey, Edmund,
  Charles Lever: His Life in His Letters, 2 vols, Blackwood & Sons, London and Edinburgh, 1906.

Dunne, Tom,
  Maria Edgeworth and the Colonial Mind, 26th O'Donnell Lecture, delivered at University College, Cork on 27th June, 1984.

Edgeworth, Maria,
  Castle Rackrent, Oxford University Press, Oxford, 1981.
  Ennui, Johnson, London, 1812–1813.
  Moral Tales, 2 vols, R. Hunter and Others, London, 1826.
  Ormond, Allen Sutton Publishing Co., Stroud, 1990.
  The Absentee, Oxford University Press, Oxford, 1988.
  The Parents' Assistant; Or, Stories for Children, 3 vols, J. Cumming, Dublin, 1829.

Edwards, R.D. and Williams, T.D. (eds.),

The Great Famine: Studies in Irish History 1845–1852, Browne & Nolan, Dublin, 1956.

Faxon, Frederic Winthrop,
Literary Annuals and Gift Books, United Reference Books, The Boston Book Company, Boston, Mass., 1912.

Flanagan, Thomas,
The Irish Novelists, 1800–1850, Columbia University Press, New York, 1959.

Forster, John,
The Life of Charles Dickens, 3 vols, Chapman & Hall, London, 1873.

Foster, R.F.,
Modern Ireland 1600–1972, Allen Lane, The Penguin Press, London, 1988.

Frith, W.P.,
My Autobiography and Reminiscences, 4 vols, Bentley, London, 1887.

Galt, John,
The Annals of the Parish and The Ayrshire Legatees, new edition, Blackwood, Edinburgh and London [n.d.]

Gilpin, William,
Observations on the River Wye and Several Parts of South Wales, &c, Relative Chiefly to Picturesque Beauty in the Summer of ... 1770, 5th edition, Cadell, London, 1800.

Observations on Several Parts of Great Britain, Particularly the Highlands of Scotland, Relative Chiefly to Picturesque Beauty, 2 vols, 3rd edition, Cadell, London, 1808.

Remarks on Forest Scenery and Other Woodland Views Relative Chiefly to Picturesque Beauty Illustrated by the Scenes of New Forest in Hampshire, 2 vols, 3rd edition, Cadell, London, 1808.

Three Essays on Picturesque Beauty; On Picturesque Travel; and on Sketching Landscape: to which is added a Poem on Landscape Painting, Blamire, London, 1794.

Goss, W.H.,
The Life and Death of Llewellyn Jewitt, F.S.A. Henry Gray, London, 1899.

Graham, Walter J.,
The Beginnings of English Literary Periodicals: A Study of Periodical Literature 1665–1715, Oxford University Press, New York, 1930.

Grant, Mrs Beatrice,
The History of An Irish Family, Miller & Haddington, Dunbar, 1822.

Gregg, Pauline,
A Social and Economic History of Britain 1760–1950, Harrap, London, 1950.

Griffin, Gerald,
Holland Tide; Or Munster Popular Tales, reprint of 1827 edition, Garland Publishing Inc., New York and London, 1979.

Tales of the Munster Festivals, 3 vols, reprint of 1827 edition, Garland Publishing Inc., New York and London, 1979.

Bibliography

The Christian Physiologist and Other Tales, James Duffy, Dublin and London [n.d.]

The Collegians, Or The Colleen Bawn: A Tale of Garryowen, 3 vols, reprint of 1829 edition, Garland Publishing Inc., New York and London, 1979.

The Rivals: Tracy's Ambition, 3 vols, reprint of 1829 edition, Garland Publishing, Inc., New York and London, 1979.

Hall, Samuel Carter, *A Memory of Thomas Moore*, Virtue, London, McGee, Dublin, 1879.

Retrospect of a Long Life, from 1815 to 1883, 2 vols, Bentley, London, 1883.

A Book of Memories of Great Men and Women of the Age, Virtue, London, 1877.

Hall, S.C. and Mrs Anna Maria Hall,
Hand-books for Ireland, 4 vols, J. McGlashan, Dublin, 1853.
Ireland, its Scenery, Character, etc., 3 vols, How & Parsons, London, 1841–43.
New edition, Virtue, London, 1853.

Hamilton, Catherine,
Notable Irishwomen, Sealy, Dublin, 1904.
Women Writers: Their Works and Ways, Ward Lock, London, 1892.

Hamilton, Elizabeth,
The Cottagers of Glenburnie, Manners & Miller, Edinburgh, 1808.

Harmon, Maurice and McHugh, Roger,
A Short History of Anglo-Irish Literature, Wolfhound Press, Dublin, 1982.

Hayley, Barbara,
'A Reading and Thinking Nation' in *300 Years of Irish Periodicals*, eds. Barbara Hayley, and Enda McKay, Association of Irish Periodical Journals, Mullingar, 1987.
A Bibliography of the Writings of William Carleton, Colin Smythe, Gerrards Cross, 1985.
Carleton's Traits and Stories *and the 19th Century Anglo-Irish Tradition*, Colin Smythe, Gerrards Cross, 1983.

Hickey, Rev. William ('Martin Doyle'), *An Address to the Landlords of Ireland, on Subjects Connected with the Melioration of the Lower Classes*, Curry, Dublin 1831.
Hints Addressed to the Smallholders and Peasantry of Ireland on Road Making and on Ventilation, etc., Curry, Dublin, 1830.
Hints Originally Intended for the Small Farmers of the County of Wexford, Ireland, Curry, Dublin, 1835.
Irish Cottagers, Curry, Dublin, 1833.

Hill, Constance, *Mary Russell Mitford and Her Surroundings*, Lane, London, 1920.

Hobsbawm, E.J. and Rudé, George,
Captain Swing, Lawrence & Wishart, London, 1969.

Home, Daniel D.,
Incidents in My Life, 2 vols, Tinsley, London, 1872.
Houfe, Simon (ed.),
The Dictionary of British Book Illustrators and Caricaturists, 1800–1914, Antique Collectors' Club, Woodbridge, Suffolk, 1978.
Houghton, R., and Slingerland, J.H. (eds.),
The Wellesley Index to Victorian Periodicals 1824–1900, Vol. IV, University of Toronto Press and Routledge & Kegan Paul, London, 1967.
Inglis, Brian,
The Freedom of the Press in Ireland 1784–1841, Faber, London, 1954.
Inglis, Henry David,
Ireland in 1834. A Journey Throughout Ireland During the Spring, Summer and Autumn of 1834, 2 vols, Whittaker, London, 1834.
James, Louis, *Fiction for the Working Man 1830–1850: A Study of the Literature Produced for the Working Classes in Early Victorian England*, Oxford University Press, 1963.
James, Louis (ed.),
Print and the People 1819–1851, Allen Lane, London, 1976.
Jeffares, A. Norman,
Anglo-Irish Literature, Gill & Macmillan, Dublin, and Macmillan, London, 1982.
Kelly, Gary,
English Fiction of the Romantic Period 1789–1830, Longman, London and New York, 1989.
Kiely, Benedict,
Poor Scholar: A Study of the Works and Days of William Carleton, Sheed & Ward, London, 1947.
Kilroy, James F. (ed.),
The Irish Short Story: A Critical History, Twayne Publishers, Boston, 1984.
Krans, Horatio S.,
Irish Life in Irish Fiction, Macmillan, New York, 1903.
Lambourne, Lionel,
An Introduction to Victorian Genre Painting from Wilkie to Frith, HMSO, London, 1982.
Leadbeater, Mary,
Cottage Dialogues Among the Irish Peasantry, Part II, Johnson, Dublin, 1813.
Tales for Cottagers, Accommodated to the Present Condition of the Irish Peasantry, J. Cumming, Dublin, 1814.
Lebow, Richard Ned,
White Britain and Black Ireland: The Influence of Stereotypes on Colonial Policy, Institute for the Study of Human Issues, Philadelphia, 1976.
Lever, Charles,
Charles O'Malley, the Irish Dragoon, 2 vols, Curry, Dublin, 1841.

Jack Hinton, the Guardsman, Curry, Dublin [nd.]
Lord Kilgobbin, George Routledge & Sons, London and New York [nd.]
The Knight of Gwynne, Chapman & Hall, London, 1847.
The O'Donoghue: A Tale of Ireland Fifty Years Ago, Curry, Dublin, 1845.
Tom Burke of 'Ours', 2 vols, Curry, Dublin, 1844.
St. Patrick's Eve, Chapman & Hall, London, 1845.

Lewis, George Cornewall,
Local Disturbances in Ireland, Tower Books, Cork, 1977.

Lover, Samuel,
Handy Andy: A Tale of Irish Life, O'Donoghue, D.J., Constable, London, 1898.

Lover, Samuel,
Legends and Stories of Ireland, Wakeman, Dublin, 1831.
Popular Tales and Legends of the Irish Peasantry, Wakeman, Dublin, 1834.
Rory O'More: A National Romance, 3 vols, Bentley, London, 1837.

MacCarthy, B.G.,
'Irish Regional Novelists of the Early Nineteenth Century', in *Dublin Magazine*, Vol. XXI, No.1, Jan–March 1946, pp. 26–32, and in Vol. XXI, N.s. No. 3, Jul–Sept. 1946, pp. 28–39.
Women Writers up to 1818, Cork University Press, 1947.

McCormack, W.J.,
Ascendancy and Tradition in Anglo-Irish Literary History from 1789 to 1939, Clarendon Press, Oxford, 1985.

MacDonagh, Oliver,
States of Mind: A Study of Anglo-Irish Conflict, 1780–1980, George Allen & Unwin, London, 1983.
The Nineteenth Century Novel and Some Aspects of Irish Social History, O'Donnell Lecture delivered at University College, Cork, 21 April, 1970. National University of Ireland, Dublin [nd.]
The Hereditary Bondsman: Daniel O'Connell 1775–1829, Weidenfeld & Nicholson, London, 1988.
The Emancipist: Daniel O'Connell, 1830–1847, Weidenfeld & Nicholson, London, 1989.

McDowell, R.B.,
Public Opinion and Government Policy in Ireland, 1801–1846, Faber, London, 1957.

McHugh, Roger, 'William Carleton: A Portrait of the Artist as Propagandist', in *Studies*, Vol. XXVI, March 1938, pp. 47–62.
'Charles Lever' in *Studies*, Vol. XXVI, June 1938, pp. 247–260.
'Maria Edgeworth's Irish Novels' in *Studies*, Vol. XXVII, Dec. 1938, pp. 556–570.

McHugh, Roger and Harmon, Maurice,
A Short History of Anglo-Irish Literature from its Origins to the Present Day, Wolfhound, Dublin, 1982.

Madden, R.R., *The History of Irish Periodical Literature from the End of the Seventeenth Century: Its Origins, Progress and Results*, 2 vols, Thom, London, 1867.

Manwaring, E.W.,
Italian Landscape in Eighteenth Century England: A Study Chiefly of the Influence of Claude Lorraine and Salvator Rosa on English Taste, 1700–1800, Oxford University Press, New York, 1925.

Marshall, Dorothy,
Industrial England 1776–1851, Routledge & Kegan Paul, London and Melbourne, 1982.

Martineau, Harriet,
Ireland, A Tale; (Illustrations of Political Economy, No. 9), Fox, London, 1832.

Maturin, Charles,
The Wild Irish Boy, 3 vols, reprint of 1808 edition. Garland Publishing Inc., London and New York, 1979.

Memmi, Albert,
The Colonizer and the Colonized, Souvenir Press, London, 1975.

Mill, John Stuart,
Collected Works, 14 vols, ed. F.E. Mineka, University of Toronto Press, Toronto, Routledge & Kegan Paul, London, 1963.

Moore, Thomas,
Memoirs, ed. Russell, 8 vols, Longman, London, 1856.
Memoirs of Captain Rock, The Celebrated Irish Chieftain, with Some Account of His Ancestors, Written by Himself, Longman, London, 1824.

More, Hannah,
Coelebs in Search of a Wife; Comprehensive Observations on Domestic Habits and Manners, etc., T. Allman, London, [nd.]

Morgan, Lady (Sydney Owenson),
Florence MaCarthy: An Irish Tale, 4 vols, reprint of 1818 edition, Garland Publishing Inc., New York and London, 1979.
Manor Sackville in Dramatic Scenes From Real Life, 3 vols, reprint of 1833 edition, Garland Publishing Inc., 1979, New York and London.
O'Donnell: A National Tale, 3 vols, reprint of 1814 edition, Garland Publishing Inc., New York and London, 1979.
The Life and Times of Salvator Rosa, 2 vols, Colburn, London, 1824.
The O'Briens and The O'Flahertys: A National Tale, 4 vols, reprint of 1827 edition, Garland Publishing Inc., New York and London, 1979.
The Wild Irish Girl, 3 vols, reprint of 1806 edition, Garland Publishing Inc, New York & London, 1979.

Moynihan, Julian,
'The Image of the City in Nineteenth Century Irish Fiction', in *The Writer and the City*, ed. Maurice Harmon, Colin Smythe, Gerrards Cross, 1984, pp. 1–17.

Newomer, James,
'Mrs. Samuel Carter Hall and *The Whiteboy*, in *Etudes Irian Daises* vol. VIII, Dec. 1983, pp. 113–119.

Nowlan, Kevin B.,
'Agrarian Unrest in Ireland, 1800–1845', in *University Review*, Vol. II, No. 5, 1959, pp. 7–16.

O'Brien, George,
Economic History of Ireland from the Union to the Famine, Longmans, London, 1921.

O'Donoghue, Denis J.,
The Life of William Carleton, being his Autobiography and Letters; and an account of his Life and Writings from the point at which his autobiography breaks off, 2 vols, Downey, London, 1895.

Osborne, Richard Boyse,
The Days of Richard Boyse Osborne from the Backwoods of Upper Canada, 1834 being of years 108 (unpublished MSS in NLI).

O'Tuathaigh, Gearóid,
Ireland Before the Famine, 1798–1848, Gill & Macmillan, Dublin, 1972.

Power, John,
List of Irish Periodical Publications (chiefly literary) from 1729 to the Present Time, Printed for private distribution only, by J. Martin. [nd.]

Rafroidi, Patrick,
'The Uses of Irish Myth in the Nineteenth Century' in *Studies*, No. 62, Dublin, 1973, pp. 257–261.

Roche, Regina Mary,
The Children of the Abbey: A Tale, Smyth, Belfast [nd.]

Ruskin, John,
Praeterita, 3 vols, Allen, Orpington and London, 1900.

Ryle, Gilbert,
Collected Papers, 2 vols, Hutchinson, London, 1971.

Sadlier, Michael,
Nineteenth Century Fiction: A Bibliographical Record based on his own Collection, 2 vols, Constable, London, 1951.

Saintsbury, George,
A History of Nineteenth Century Literature 1780–1895, Macmillan, New York, 1896.

Sanders, Andrew,
The Victorian Historical Novel 1840–1880, Macmillan, London, 1978.

Sloan, Barry,
The Pioneers of Anglo-Irish Fiction, 1800–1850, Colin Smythe, Gerrards Cross, 1986; Barnes & Noble Books, Totowa, N.J., 1987.

Sullivan, Alvin (ed.),
British Literary Magazines, vol. II, *The Romantic Age 1789–1836*,

Greenwood Press, Connecticut and London, 1983.
Tillotson, Kathleen,
Novels of the Eighteen Forties, Clarendon Press, Oxford, 1954.
Tomkins, J.S.,
The Popular Novel in England 1700–1800, Constable, London, 1932.
Tonna, Charlotte Elizabeth ('Charlotte Elizabeth'),
Derry; A Tale of the Revolution, 5 edition, Nisbet, Dublin, 1837.
Letters from Ireland mdcccxxxvii, Seeley, Dublin, 1838.
Trench, W. Stewart,
Realities of Irish Life, Longmans, Green & Co., London, 1868.
Trevelyan, Charles,
The Irish Crisis, Longman, London, 1848.
Twiss, Richard,
A Tour in Ireland in 1775, 2nd edition, Sheppard, Dublin, 1776.
Vaughan, W.E.,
Landlords and Tenants in Ireland 1848–1904, Studies in Irish Economics and Social History, No. 2, Dundalk, 1984.
Vizetelly, Henry,
Glances Back Through 70 Years; Autobiographical and Other Reminiscences, 2 vols, Kegan Paul, Trench, Trübner & Co. Ltd., London, 1893.
Wall, Maureen,
'The Whiteboys', in *Secret Societies in Ireland*, ed. T. Desmond, Williams, Gill & Macmillan, 1973, pp. 13–25.
Wheeler, Michael,
The Art of Allusion in the Victorian Novel, Macmillan, London, 1979.
Whelan, Kevin (ed.),
Wexford, History and Society, Geography Publications, Dublin, 1987.
White, Cynthia L.,
Women's Magazines, 1693–1968, Michael Joseph, London, 1970.
Wilson, John,
Lights and Shadows of Scottish Life in *The Works of Professor Wilson of the University of Edinburgh*, ed. Professor Ferrier, vol. XI, Blackwoods, Edinburgh and London, 1855–58.
Winstanley, Michael J.,
Ireland and the Land Question 1800–1922, Methuen, London, 1984.

(D) PERIODICALS

NLI : National Library of Ireland
BL : British Library
COL : British Library Newspaper Library, Colindale

Amulet, London, 1826–1836, (BL)
Art Union, London, 1841–1863.

Atlas, London, May 1826–December 1852, (COL)
Belfast Newsletter, Belfast, 1823–1835, 1837–1852, (NLI)
Bentley's Miscellany, London, 1837–1852, (BL)
Blackwood's Edinburgh Magazine, Edinburgh, 1817–1852, (BL, NLI, TCD)
Catholic Penny Magazine, Dublin, February 1843–April 1835 (NLI)
Chambers's Journal, Edinburgh, 1832–1852, (BL, NLI)
Christian Examiner and Church of Ireland Magazine, Dublin, 1825–1835, (NLI, TCD)
Citizen; A Monthly Journal of Politics, Literature and Art, Vols I–II, 1839–1840; afterwards The Citizen; or, Dublin Monthly Magazine, Vols 3–4, 1840–1841; Dublin Monthly Magazine; being a new series of the Citizen and Including the Native Music of Ireland, January–December 1842; and the Dublin Magazine (and Citizen), Dublin, January–April 1843, (NLI)
Dublin Evening Mail, Dublin, 1823–1852, (NLI)
Dublin Monitor, Dublin, 1838–1845, (NLI)
Dublin Literary Gazette or, Weekly Chronicle of Criticism, Belles Lettres and Fine Arts, Dublin, January–June 1830. Afterwards the National Magazine, July–December, 1830; and the National Magazine and Dublin Literary Gazette, Dublin, January–April, 1831 (NLI)
Dublin Monthly Magazine, Dublin, January–June 1830, (NLI, TCD)
Dublin Penny Journal, Dublin, June 1832–June 1836, (NLI)
Dublin University Magazine, Dublin, 1833–1852, (NLI, TCD)
Duffy's Irish Catholic Magazine, Dublin, February 1847, December, 1848, (NLI)
Eclectic Review, London, 1805–1852, (BL, NLI, TCD)
Edinburgh Literary Gazette, Edinburgh, Nos. 16–62, 1823–24, (BL)
Edinburgh Literary Journal, Edinburgh, 1828–1831, (BL, NLI, TCD)
Edinburgh Review or Critical Journal, Edinburgh, 1803–1852, (BL, NLI, TCD)
Farmer's Gazette and Journal of Practical Horticulture, Dublin 1843–1852, (NLI, TCD)
Fraser's Magazine for Town and Country, London 1830–1852, (BL, NLI, TCD)
Gentleman's Magazine or Monthly Intelligencer, London, 1750–1833, N.S. 1834–1856, (BL, NLI, TCD)
Howitt's Journal of Literature and Popular Progress, London, 1847, (BL)
Irish Farmer's and Gardener's Magazine and Register of Rural Affairs, Dublin, November 1833–December 1840 (NLI)
Irish Farmers' and Gardeners' Register, Dublin, March 1842, February 1843, (NLI)
Irish Farmers' Journal, Dublin, 5 August–31 December 1845, (NLI)
Irish Farmers' Journal and Weekly Intelligencer, Dublin, September 1812–February 1827, (NLI)
Irish Monthly Magazine of Politics and Literature, Dublin, May 1832–September 1834, (NLI, TCD)
Irish Penny Magazine, Dublin, January 1833–January 1834, (NLI)

Irish Union Magazine, Dublin, 1845, afterwards the *Irish Monthly Magazine*, 1842–1846 (NLI)
Irish Penny Journal, Dublin, July 1840–June 1841, (NLI)
John Bull, London, 1820–1852, (COL)
Lady's Magazine, 1770–1832, continued as the *Lady's Magazine and Museum of Belles Lettres*, London, 1832–1837, continued as the *Court Magazine and Monthly Critic and Ladies' Magazine and Museum of The Belles Lettres*, London, 1838, (BL, NLI)
Literary Gazette and Journal of Belles Lettres, Arts, Sciences, etc., London, 1817–1852, (BL, NLI, TCD)
Metropolitan Magazine, London, 1833–1850 (BL)
Metropolitan, London, 1831–1832 (BL)
Monthly Chronicle, London, 1838–1841, (BL)
Monthly Magazine and British Register, London, 1796–1826, continued as the *Monthly Magazine; or British Register of Literature, Sciences and The Belles Lettres*, London, 1826–1834, continued as the *Monthly Magazine of Politics, Literature and The Belles Lettres*, 1835–1838, continued as the *Monthly Magazine*, London, 1839–1843 (BL)
Monthly Review, London, 1826–1828, 1828–1830, 1831–1845, (BL)
Munster Farmer's Magazine, Cork, 1812–1814 (NLI)
Nation, Dublin, October 1842–January 1848 (NLI)
New Monthly Magazine and Universal Register, 1814–1820, continued as *New Monthly Magazine and Literary Journal*, 1821–1836, continued as *New Monthly Magazine and Humorist*, London, 1837–1852, (BL, NLI, TCD)
North British Review, Edinburgh, 1844–1852, (BL)
Northern Whig, Belfast, 1 January 1824–30 April 1829, January 1829–December 1850, (NLI)
Observer, London, 1829–1852 (COL)
Penny Magazine, Dublin 1832–1844, (NLI)
Prospective Review, London, 1844–1855, (BL)
Punch, or The London Charivari, 1845–1850, (BL)
Quarterly Review, London, 1809–1852, (BL, NLI, TCD)
Scots Magazine, Edinburgh, 1760–1803, (BL)
Sharpes' London Magazine, London, 1845–1849, continued as *Sharpes' London Journal*, March–October, 1849, (BL, NLI, TCD)
Spectator, London, 1800–1852, (BL, NLI, TCD)
The Sunday Times, London, 1829–1852, (COL)
Tait's Edinburgh Magazine, Edinburgh, 1832–1852, (BL, NLI, TCD)
Tablet, London, May 1840–July 1842, January 1843–December 1850 (NLI)
Tatler, London, 1830–1831, N.S. April–October 1832 (BL)
The Times, London, 1829–1852, (COL)
University Review and Quarterly Magazine, afterwards the *Dublin University*

Review and Quarterly Magazine, Dublin, January and April, 1833, (NLI, TCD)
Warder, Dublin November 1832–December 1850, (NLI)
Weekly Dispatch, London, 1817–1852, (COL)
Westminster Review, London 1824–1836, continued as *London and Westminster Review*, London, 1836–1840, continued as *Westminster Review*, London, 1836–1841, continued as *Westminster Review*, London, 1841–1852, (BL, NLI, TCD)
Wexford Conservative, Wexford, September 1832–April 1846, (NLI)
Wexford Evening Post, Wexford, March 1826–March 1830 (NLI)
Wexford Freeman, Wexford, May 1832–May 1837, (NLI)
Wexford Herald, Wexford, 2 January, 1828–31 August, 1832, (NLI)

(E) THESES

Boué, André,
  William Carleton 1794–1869, Romancier Irlandais, Paris, Publications de la Sorbonne, 1978.
Cohane, Mary Ellen,
  Style in the Novels of William Carleton: A Dissertation on Folklore and Folklife. Ph.D. dissertation, University of Pennsylvania, 1984.
Harrison, S.J.C.,
  Irish Women Writers, 1800–1835, Ph.D. thesis, Trinity College, Dublin, 1947.
Hayley, Barbara,
  William Carleton in Text and Context – a comparative study of Traits and Stories of the Irish Peasantry, 2 vols, Ph.D. thesis, Canterbury, University of Kent, 1977.
Ibarra, Eileen S.,
  Realistic accounts of the Irish Peasantry in four novels of William Carleton, Ph.D. thesis, University of Florida, 1969.
McBride, John P.,
  The Dublin University Magazine: *Cultural Nationality and Tory Ideology in an Irish Literary and Political Journal 1833–1852*, 2 vols, Ph.D. thesis, Trinity College, University of Dublin, 1987.
Meredith, Robert L.,
  Charles Lever: Anglo-Irish Novelist, Ph.D. thesis, Duke University, 1975.

NOTES

1. *Sketches of Irish Character*, 2 vols, Westley & Davis, London, 1829.
2. Mrs Hall, *Sketches of Irish Character*, Second Series, 2 vols, Westley & Davis, London, 1831.
3. Mrs Hall, *Lights and Shadows of Irish Life*, 3 vols, Colburn, London, 1838.
4. Mrs Hall, *Stories of the Irish Peasantry*, Chambers, Edinburgh, 1840.
5. Mr and Mrs S.C. Hall, *Ireland; its Scenery, Character, etc.*, 3 vols, How & Parsons, London, 1843.
6. Mrs Hall, *The Whiteboy; A Story of Ireland in 1822*, 2 vols, Chapman & Hall, London, 1845.
7. Mrs Hall, *Midsummer Eve; A Fairy Tale of Love*, Longman, London, 1847.
8. Mrs Hall, *The Fight of Faith*, Chapman & Hall, London, 1869.

CHAPTER 1
1. Elizabeth Manwaring, *Italian Landscape in Eighteenth-Century England; A Study Chiefly of the Influence of Claude Lorraine and Salvator Rosa on English Taste*, The Wellesley Semi-Centennial Series, New York, 1925, p. 23
2. Mrs S.C. Hall, *Sketches of Irish Character*, Second Series, 2 vols, Westley & Davis, London, 1831.
3. Mrs S.C. Hall, 'The Rapparee', *Sketches of Irish Character*, Second Series, p. 127.
4. Mrs S.C. Hall, *The Whiteboy; A Story of Ireland in 1822*, 2 vols, Chapman & Hall, London, 1845.
5. *Ibid.* II, p. 12.
6. Mr & Mrs S.C. Hall, *Ireland; its Scenery, Character, etc.*, 3 vols, How & Parsons, London, 1843.
7. *Eclectic Review*, Vol VI, Part I, May 1810, p. 417.
8. Mrs S.C. Hall, *Sketches of Irish Character*, First Series, 2 vols, Westley & Davis, London, 1829.
9. Mrs S.C. Hall, *Sketches of Irish Character*, Fifth Edition, London, 1854, Introduction, p. xvi.
10. Mrs S.C. Hall, *Sketches of Irish Character*, Third Edition, How & Parsons, London, 1842, Introduction, pp. v, vi.

11 Mr S.C. Hall, *Retrospect of a Long Life, from 1815 to 1883*, 2 vols, Bentley, London, 1883, pp. 422–25.
12 Richard Boyse Osborne, *Diary, 1834–1886*, Unpublished Manuscript in NLI MSS 7888, pp. 4–5.
13 Mrs S.C. Hall, Dedication, *Sketches of Irish Character*, Third Edition, How & Parsons, London, 1842.
14 Father Thomas Butler O.S.A., *A Parish and Its People*, Wexford, 1985, p. 174.
15 Thomas Moore, *Memoirs, Journal and Correspondence*, ed. Lord John Russell, 8 vols, London, Longmans, 1856, VII, pp. 110–13.
16 *Literary Gazette*, No. 640, 25 April 1829, pp. 268–69.
17 *New Monthly Magazine*, Vol. XXVIII, No. 102, June 1829, pp. 241–42.
18 *Eclectic Review*, Vol. II, No. 50, July 1829, p. 72.
19 Allan Cunningham, 'British Novels and Romances', Nov. 1833, pp 809–15; *Athenaeum*, Nov. 1833, No. 318, p. 812.
20 *Fraser's Magazine for Town and Country*, Vol. IV, No. 19, Aug. 1831, pp. 101–12.
21 Mrs S.C. Hall, *Chronicles of a Schoolroom*, Westley & Davis, London, 1831.
22 *Edinburgh Literary Journal* No. 76, 24 April 1830, p. 244.
23 Mrs S.C. Hall, *Tales of Woman's Trials*, Houlston, London, 1835.
24 *Literary Gazette*, No. 934, 29 Nov. 1834, pp. 796–97.
25 Mrs S.C. Hall, *Lights and Shadows of Irish Life*, 3 vols, Colburn, London, 1838.
26 *The Sunday Times*, No. 809, 20 April 1838, p. 3.
27 *Ibid*, No. 810, 29 April 1838, p. 7.
28 *Observer*, 6 May 1838, p. 3.
29 *Spectator*, Vol. II, No. 513, 28 April 1838, p. 398.
30 *John Bull*, Vol. XVIII, No. 907, p. 201.
31 *Weekly Dispatch*, No. 1906, 29 April 1838, p. 202.
32 Mrs S.C. Hall, *Stories of the Irish Peasantry*, Chambers, Edinburgh, 1840.
33 Mrs S.C. Hall, *Marian; Or a Young Maid's Fortunes*, 3 vols, Colburn, London, 1840.
34 *Monthly Chronicle*, Vol. V, April 1840, p. 222.
35 *Athenaeum*, No. 644, 29 Feb. 1841, p. 171.
36 *New Monthly Magazine*, Vol. LVIII, No. 230, Feb. 1840, p. 287.
37 *The Sunday Times*, No. 1307, 7 Nov. 1847, p. 3.
38 Mr and Mrs S.C. Hall, *Ireland; its Scenery, Character, etc.*, 3 vols, How & Parsons, London, 1843.
39 Mrs S.C. Hall, *The Whiteboy; A Story of Ireland in 1822*, 2 vols, Chapman & Hall, London, 1845.
40 *Fraser's Magazine*, Vol. XXIII, No. 113, Jan. 1841, pp. 90–4.
41 *The Sunday Times*, No. 1055, 8 Jan. 1843, p. 2.

42 *Ibid.*, No. 1092, 24 Sept. 1843, p. 2.
43 *Gentleman's Magazine*, Vol. XXIV, Nov. 1845, p. 507.
44 *John Bull*, Vol. XXV, No. 1,288, 16 Aug. 1845, p. 520.
45 *Atlas*, Vol. XX, No. 1,004, 9 Aug. 1845, p. 507.
46 Mrs S.C. Hall, *Midsummer Eve; A Fairy Tale of Love*, Longman, London, 1847.
47 *Observer*, 12 Dec. 1847, p. 7.
48 *Atlas*, Vol. XXII, No. 1,126, 11 Dec. 1847, p. 831.
49 Mrs S.C. Hall, *The Fight of Faith*, 2 vols, Chapman & Hall, London, 1869.
50 *Fraser's Magazine*, Vol. II, No. IC, Oct. 1830, pp. 312–13.
51 *Edinburgh Review*, Vol. XLIII, No. 86, Art V, 26 Feb. 1826, p. 356.
52 *Edinburgh Literary Gazette*, Vol. 1, No. 8, 4 July 1829, p. 116.
53 *Monthly Magazine*, Vol. VIII, No. 45, Sept. 1829, p. 334.
54 *Edinburgh Literary Journal*, No. 29, 30 May 1829, pp. 405–06.
55 *Ibid.*, No. 129, 30 Apr. 1831, p. 278.
56 *Athenaeum*, No. 552, 1838, p. 375.
57 *Gentleman's Magazine*, Vol. XXIV, Nov. 1845, p. 507.
58 *New Monthly Magazine*, Vol. LXXV, No. 297, 1845.

CHAPTER II
1 Mrs S.C. Hall, *Retrospect of a Long Life*, 1815–1883, 2 vols, Bentley, London, 1889. (Hereinafter referred to as *Retrospect*).
2 *Ibid.*, I, p. 100.
3 Richard Boyse Osborne, *Diary 1834–1886* (Unpublished Manuscript in NLI, MSS 7888 pp. 3, 4.
4 Mrs S.C. Hall, II, *Retrospect* p. 425 f.n.
5 *The Spirit and Manners of the Age*, Vol. I, No. 1, Jan. 1826, Introduction, p. 1.
6 Mrs S.C. Hall, Introduction, *Sketches of Irish Character*, Fifth Edition, Hotten, London, 1854, pp. x, xi.
7 *Ibid.*, p. xii.
8 *Spirit and Manners of the Age*, Vol. 2, N.S. May 1829, p. 385.
9 Mrs S.C. Hall, Introduction, *Sketches of Irish Character*, Fifth Edition, p. xii.
10 Mrs S.C. Hall, *Boons and Blessings*, Virtue, Spalding & Co., London, 1875, p. 98.
11 Mrs S.C. Hall, II, f.n. *Retrospect* p. 486.
12 Daniel D. Home, *Incidents in My Life*, 2 vols, Tinsley, London, 1872, I, p. 29.
13 *Ibid.*, I, pp. 121–122.
14 Mrs S.C. Hall, *The Use of Spiritualism*, Hay, Nisbet & Co., Glasgow, Allen, London, 1884, pp. 76–7.
15 Mrs S.C. Hall, 'The Naughty Boy', *The Village Garland, Tales and Sketches*, Chapman & Hall, London, 1863, pp. 252–3.
16 Mrs S.C. Hall, 'Dedication', *The Juvenile Budget*, Chapman & Hall,

London, 1837, p. vii.
17 Mrs S.C. Hall, Letter of Nov. 23, no year, MS 17064, NLI.
18 Mrs S.C. Hall, Introduction, *Sketches of Irish Character*, Fifth Edition, pp. XIX, XX.
19 Mrs S.C. Hall, *The Juvenile Budget*, pp. 65–6.
20 Mrs S.C. Hall, II, *Retrospect* p. 446.
21 Preface, The *Amulet*, Jan. 1827, p. ii.
22 *Lady's Magazine*, Vol. VIII, Oct. 1827, p. 554.
23 *Ibid.*, NS, Nov. 30, 1828, p. 571.
24 *Ibid.*, p. 590.
25 *Blackwood's Magazine*, Vol. XXVI, No. 160, Dec. 1829, p. 974.
26 Preface, *Amulet*, 1832, p. iii.
27 Frederic Winthrop Faxon, *Literary Annuals and Gift Books*, United Reference Series, The Boston Book Company, Boston, Mass, 1912, p. xxi.
28 John Ruskin, *Praeterita*, 3 vols, Allen, Orpington and London, 1900, I, pp. 148–49.
29 *Ibid.*, p. 150.
30 *Atlas*, Vol. VI, No. 285, 30 Oct. 1831, p. 732.
31 *Fraser's Magazine*, Vol. XV, No. 85, Jan. 1837, pp. 37–8.
32 *Dublin Penny Journal*, 1835–36, p. 245.
33 John Stuart Mill, *Collected Works*, 14 vols, Vol. XIII, *The Earlier Letters of John Stuart Mill*, ed. Francis E. Mineka, University of Toronto Press, Toronto, Routledge & Kegan Paul, London, 1963. Letter 262 to John Robertson, Sept. 1839, p. 403.
34 Mrs S.C. Hall, *Lights and Shadows of Irish Life*, 3 vols, Henry Colburn, London, 1838.
35 Barbara Hayley, 'Nineteenth Century Irish Periodicals', *Three Hundred Years of Irish Periodicals*, eds. Barbara Hayley and Enda McKay, Lilliput Press, Mullingar, 1987 pp. 29–46. (Hereinafter referred to as *Irish Periodicals*).
36 *Dublin Literary Gazette and Weekly Chronicle*, Vol. I, No.1, Jan. 1830, p. 2.
37 Hayley, *Irish Periodicals*, p. 34.

CHAPTER III
1 A. Norman Jeffares, 'Anglo-Irish Literature', *Macmillan History of Literature*, Macmillan, London, Gill & Macmillan, Dublin 1982, p. 132.
2 Henry Fielding, Dedication, *The History of Tom Jones; A Foundling*, Penguin Classics, Penguin Books, London, 1987, p. 37.
3 Samuel Richardson, Preface, *Pamela; Or, Virtue Rewarded*, Penguin Classics, Penguin Books, London, 1985, p. 31.
4 Samuel Richardson, Preface, *Clarissa; Or, the History of a Young Lady*, Penguin Classics, Penguin Books, London, 1985, p. 36.
5 John Wesley, Preface, Henry Brooke, *The Fool of Quality*, abridged by

John Wesley, Tegg, London, 1780, p. vi.
6. *Edinburgh Review*, Vol. XIV, No. 27, Apr. 1809, pp. 145–6.
7. Angus Ross, 'Hannah More', *The Penguin Companion to Literature*, 4 vols, Penguin Books, London, 1971, 1, p. 376.
8. B.G. MacCarthy, *Women Writers up to 1818*, Cork University Press, Cork, 1947, p. 70.
9. Maria Edgeworth, *Chosen Letters*, edited by F.V. Barry, Cape, London, 1931, p. 213.
10. Gilbert Ryle, 'Jane Austen and the Moralists', *Collected Papers*, 2 vols, Hutchinson, London, 1971, 11, pp. 266–90.
11. 'Sunday Reading', *Irish Farmers' Journal and Weekly Intelligencer*, Vol. I, No. 47, Jul. 1813, pp. 390–91.
12. *Eclectic Review*, Vol. II, Part I, Feb. 1806, p. 140.
13. *Ibid.*, Vol. XI, N.S., May 1819, pp. 484–85.
14. *Gentleman's Magazine*, Vol. XCVIII, Part 2, 1828, p. 536.
15. *Athenaeum*, No. 195, 23 Jul. 1831, p. 468.
16. 'Fashionable Novels', *Monthly Magazine*, Vol. II, No. 7, July 1826, pp. 154–63 (p. 155).
17. 'Miss Edgeworth as a Philosophical Writer', *Dublin Monthly Magazine*, Vol. No. 5, May 1830, pp. 417–26 (p. 417).
18. *Christian Examiner*, Vol. X, No. 56, Feb. 1830, p. 150.
19. 'Novel Writing', *Blackwood's Edinburgh Magazine*, Vol. IV, No. 22, Jan. 1819, pp. 394–96 (p. 394).
20. 'Secondary Scottish Novels', *Edinburgh Review*, Vol. XXXIX, No. 77, Oct. 1823, pp. 158–96 (p. 196).
21. 'The Novels of Zschokke', *Tait's Edinburgh Magazine*, Vol. XII, No. 139, Jul. 1845, pp. 435–40 (p. 435).
22. *New Monthly Magazine*, Vol. LVII, No. 126, Oct. 1839, pp. 285–86.
23. *Ibid.*, Vol. LI, No. 3, Sept. 1837, pp. 125–44 (p. 141).
24. *Gentleman's Magazine*, Vol. LXXIX, Oct. 1809, p. 937.
25. *Eclectic Review*, Vol. IV, Aug. 1815, p. 159.
26. 'Miss Edgeworth's Tales of Fashionable Life', *Edinburgh Review*, Vol. XX, No. 39, July 1812, pp. 100–26 (p. 100).
27. *Blackwood's Edinburgh Magazine*, Vol. XVII, No. 100, May 1825, p. 518.
28. *New Monthly Magazine*, Vol. XIII, No. 52, Apr. 1825, p. 322 (pp. 321–34).
29. *Edinburgh Literary Gazette*, No. 29, Aug. 1823, p. 209.
30. 'Miss Martineau's Monthly Novels', *Quarterly Review*, Vol. XLIX, No. 97, Apr. 1833, pp. 136–52 (p. 150).
31. *Fraser's Magazine for Town and Country*, Vol. XVIII, No. 167, Nov. 1843, pp. 505–25 (p. 506).
32. 'Memoir of Maria Edgeworth', *Bentley's Miscellany*, Vol. XXIV, 1848, pp. 477–83 (p. 480).
33. *Athenaeum*, No. 316, 16 Nov. 1833, p. 775.

34 'Miss Edgeworth's Tales', *Edinburgh Review*, Vol. XXVIII, No. 56, Aug. 1817, pp. 390–418 (p. 391).
35 'Modern Novels', *Quarterly Review*, Vol. XXIV, Jan. 1821, pp. 352–76 (p. 353).
36 *New Monthly Magazine*, Vol. III, No. 3, Mar. 1821, p. 132.
37 *Monthly Review*, Vol. VI, No. 28, Dec. 1827, pp. 508–09.
38 *Quarterly Review*, Vol. XLIX, No. 97, Apr. 1833, p. 136.
39 'Deerbrook' *Blackwood's Edinburgh Magazine*, Vol. XLVII, No. 292, Feb. 1840, pp. 177–88 (p. 177).
40 'Recent Novels', *Fraser's Magazine*, Vol. XXXVIII, No. 233, July 1848, pp. 33–40 (p. 40).
41 *Westminster Review*, Vol. LI, No. 1, Apr. 1849, pp. 48–63.
42 *Ibid.*, 'The Contemporary Literature of England', Vol. I, NS, No. 91, Dec. 1852, pp. 247–88 (284).
43 'Fruits of the Season', *Fraser's Magazine*, Vol. XLIII, No. 248, Aug. 1850, pp. 203–14 (p. 207).
44 *Ibid.*, Vol. XLVI, No. 276, Dec. 1852, p. 633.
45 *Ibid.*, 'English Novels', Vol. XLIV, No. 262, Oct. 1851, pp. 375–91 (p. 380).
46 'The Novel and the Drama', *Blackwood's*, Vol. LVII, No. 356, June 1845, pp. 679–87 (p. 684).
47 'New and Cheap Forms of Popular Literature', *Eclectic Review*, Vol. XVIII, No. 82, July 1845, pp. 72–6 (p. 73).
48 *Athenaeum*, No. 579, 1 Dec. 1838, p. 855.
49 Preface, p. 1, *North British Review*, Vol. 1, No. 1, May 1844.
50 *Ibid.*, 'Recent Works of Fiction', Vol. XV, No. 30, Aug. 1851, pp. 419–41 (p. 424).
51 'Polemical Fiction', *Prospective Review*, Vol. II, No. 28, 1851, pp. 404–23 (pp. 412–13).
52 Preface, William Carleton, *Art Maguire; Or, The Broken Pledge*, Duffy, Dublin, 1845, p. x.
53 Sir Samuel Ferguson, 'The Didactic Irish Novelists', *Dublin University Magazine*, Vol. XXVI, No. 156, Dec. 1845, pp. 737–52 (p. 737).
54 *Gentleman's Magazine*, Vol. XCII, Part II, Dec. 1822, p. 525.
55 *New Monthly Magazine*, Vol. XVIII, No. 71, Nov. 1826, p. 449.
56 'Memoirs of Captain Rock', *Westminster Review*, Vol. I, No.2, Apr. 1824, pp. 492–504 (p. 494).
57 *Westminster Review*, Vol. VII, No. 2, Apr. 1827, p. 444.
58 *New Monthly Magazine*, Vol. XX, No. 84, Dec. 1827, p. 505.

CHAPTER IV
1 Mrs S.C. Hall, 'Larry Moore', *Sketches of Irish Character*, Second Series, Westley & Davis, London, 1831, p. 333.

Notes

2 'Jack the Shrimp', *Sketches*, Second Series, p. 230.
3 'Father Mike', *Sketches*, First Series, 2 vols, Westley & Davis, London, 1829, II, p. 61.
4 'Lilly O'Brien', *Sketches*, First Series, I, p. 13.
5 'Kelly the Piper', *Sketches*, First Series, I, p. 93.
6 'Peter the Prophet', *Sketches*, First Series, II, p. 199.
7 'Mabel O'Neill's Curse', *Sketches*, Second Series, p. 26.
8 'Norah Clary's Wise Thought', *Sketches*, Second Series, p. 180.
9 *Literary Gazette*, No. 640, Apr. 1829, pp. 268–69.
10 Mrs S.C. Hall, 'Lilly O'Brien, *Sketches*, First Series, I, p. 10.
11 'The Bannow Postman', *Sketches*, First Series, II, p. 9.
12 'Hospitality', *Sketches*, First Series, II, p. 151.
13 Mrs S.C. Hall, 'Mabel O'Neill's Curse', *Sketches*, Second Series, p. 28.
14 'Father Mike', *Sketches*, First Series, II, p. 62.
15 'Norah Clary's Wise Thought', *Sketches*, Second Series, p. 183.
16 'Lilly O'Brien', *Sketches*, First Series, pp. 6–7.
17 'Kelly the Piper', *Sketches*, First Series, I, p. 93.
18 'Luke O'Brian', *Sketches*, Second Series, p. 322.
19 'We'll See About It', *Sketches*, Second Series, p. 213.
20 *Ibid.*, p. 212.
21 Ned Lebow, 'British Images of Poverty in Pre-Famine Ireland', *Views of the Irish Peasantry*, ed. Daniel J. Casey & Robert Rhodes, Hamden (Conn.), Archon, 1977, p. 59.
22 Mrs S.C. Hall, 'Larry Moore', *Sketches*, Second Series, p. 334.
23 'Independence,' *Sketches*, First Series, I, pp. 1–2.
24 'Kate Connor', *Sketches*, Second Series, p. 205.
25 'Annie Leslie', *Sketches*, Second Series, p. 205.
26 *Atlas*, No. 181, 1 Nov. 1829, p. 715.
27 *Literary Gazette*, No. 744, April 1831, p. 258.
28 *Weekly Dispatch*, No. 1559, May 1831, p. 142.
29 Mrs S.C. Hall, 'Mary Ryan's Daughter', *Sketches*, Third Edition, How & Parsons, London, 1842, p. 37.
30 'Kate Connor', *Sketches*, Second Series, p. 206.
31 Mrs S.C. Hall, 'Old Frank', *Sketches*, First Series, I, p. 215.
32 James Hogg, 'Noctes Ambrosianae', No. 41, *Blackwood's Edinburgh Magazine*, Vol. XXV, No. 150, Mar. 1829, p. 380.
33 *The Times*, 17 Aug. 1829, p. 4.
34 *Eclectic Review*, Vol. II, No. 50, July 1829, p. 72.
35 *New Monthly Magazine*, Vol. XVIII, No. 3, June 1831, pp. 247–49.
36 *Literary Gazette*, No. 744, 1831, pp. 258–59.
37 *Dublin Literary Gazette and National Magazine*, Vol. I, No. 2, Aug. 1830, p. 155.
38 E.J. Hobsbawm and George Rudé, *Captain Swing*, Lawrence & Wishart,

London, 1969, p. 81.
39. Christopher North 'Noctes Ambrosianae', *Blackwood's Edinburgh Magazine*, Oct. 1826, quoted by Constance Hall in *Mary Russell Mitford and Her Surroundings*, Lane, London, 1920, p. 210.
40. *Gentleman's Magazine*, Vol. XCIX, Part II, Oct. 1830, p. 348.
41. *Athenaeum*, No. 255, Sept. 1832, p. 593.
42. Allan Cunningham, 'Biographical and Critical History of the Literature of the Last Fifty Years', *Athenaeum*, No. 318, Nov. 1833, pp. 809–15.
43. *Blackwood's Edinburgh Magazine*, Vol. LXXI, No. 337, Mar. 1852, p. 259.
44. *Lady's Magazine*, Vol. IX, Nov. 1828, p. 587.
45. *Athenaeum*, No. 50, 1828, p. 794.
46. Mrs S.C. Hall, 'The Mosspits', *Amulet, Or Christian and Literary Remembrancer*, 1832, pp. 91–146.
47. Mrs Hall, 'Captain Andy', *Sketches*, First Series, I, p. 140.
48. 'Father Mike', *Sketches*, First Series, II, pp. 89–90.
49. 'Black Dennis', *Sketches*, First Series, I, p. 189.
50. 'The Rapparee', *Sketches*, Second Series, p. 126.
51. 'Jack the Shrimp', *Sketches*, Second Series, p. 235.
52. *Athenaeum*, No. 182, Apr. 1831, p. 262.
53. Mrs Hall, 'Captain Andy', *Sketches*, First Series, I, pp. 133–34.
54. 'Father Mike', *Ibid.*, II, pp. 91–2.
55. 'Lilly O'Brien', *Ibid.*, I, p. 24.
56. 'The Bannow Postman', *Ibid.*, II, p. 28.
57. 'Lilly O'Brien', *Ibid.*, p. 75.
58. 'The Rapparee', *Sketches*, Second Series, p. 130.
59. 'Annie Leslie', *Ibid.*, p. 100.
60. 'Kate Connor', *Ibid.*, p. 204.
61. 'The Bannow Postman', *Sketches*, First Series, II, p. 26.
62. 'Mabel O'Neill's Curse', *Sketches*, Second Series, p. 30.
63. Mrs Hall, 'Mary Ryan's Daughter', *Sketches*, Third Edition, p. 50.
64. *Ibid.*, p.50.
65. 'Annie Leslie', *Sketches*, Second Series, p. 84.
66. 'Lilly O'Brien', *Sketches*, First Series, I, p. 7.
67. 'The Wooing and the Wedding', *Sketches*, Second Series, p. 306.
68. *Ibid.*, p. 310.

CHAPTER V
1. Alison Adburgham, *Silver Fork Society*, Constable, London, 1983, p. 25.
2. Mrs S.C. Hall, *Lights and Shadows of Irish Life*, 3 vols, Colburn, London, 1838.
3. *Ibid.*, Introduction, p. vi.
4. *Ibid.*, p. vii.
5. 'The Groves of Blarney', *Lights and Shadows of Irish Life*, I, p. 5.

6   *Ibid.* I, p. 32.
7   *Ibid.* I, pp. 150–51.
8   *Ibid.* I, p. 67.
9   *Ibid.* I, p. 71.
10 *Spectator*, Vol. II, No. 513, 28 Apr. 1838, p. 398.
11 Mrs S.C. Hall, 'The Groves of Blarney', *Lights and Shadows*, I, p. 68.
12 *Athenaeum*, No. 552, 26 May, 1838, p. 375.
13 Mrs S.C. Hall, 'The Groves of Blarney', *Lights and Shadows*, I, p. 222.
14 *Ibid.*, p. 208.
15 *Ibid.*, p. 299.
16 *Dublin University Magazine*, Vol. XII, No. 68, Aug. 1838, p. 225.
17 Mrs S.C. Hall, 'Harry O'Reardon, or Illustrations of Irish Pride', *Lights and Shadows*, III, p. 6.
18 *Ibid.*, p. 7.
19 'Procrastination', *Lights and Shadows* II, p. 324.
20 'The Groves of Blarney', *Lights and Shadows*, I, p. 157.
21 *Ibid.*, p. 306.
22 'Independence', *Sketches of Irish Character*, First Series, I, p. 177.
23 Harry O'Reardon', *Lights and Shadows*, III, p. 7.
24 Father Thomas C. Butler, O.S.A., 'Big Houses and their Owners', *A Parish and Its People, History of the Parish of Carrig in Bannow*, Wexford 1985, p. 171.
25 Mrs S.C. Hall, 'The Dispensation', *Lights and Shadows*, III, p. 265.
26 *Ibid.*, p. 271.
27 *John Bull*, Vol. XVIII, No. 907, 29 Apr. 1838, p. 201.
28 Mrs S.C. Hall, 'The Dispensation', *Lights and Shadows*, III, p. 275.
29 *Ibid.*, p. 308.
30 'Ruins' Part I, *Lights and Shadows*, II, p. 195.
31 'Dermot O'Dwyer, *Lights and Shadows*, II, Luck, Part II, p. 299.
32 'The Last in the Lease', Naturals, *Lights and Shadows*, II, p. 143.
33 Preface, *Lights and Shadows*, Vol. II.
34 Mr S.C. Hall, *Retrospect of a Long Life from 1815 to 1883*, 2 vols, Bentley, London, 1883, I, p. 50.
35 Mrs S.C. Hall, 'Beggars', *Lights and Shadows*, II, p. 43, f.n.
36 Mrs S.C. Hall, 'The Last in the Lease', Naturals, *Lights and Shadows*, II, p. 159, f.n.
37 Ruins, Part II, *Lights and Shadows*, II, p. 204.
38 Ruins, Part I, *Lights and Shadows*, II, pp. 188–89.
39 Dedication, *Lights and Shadows*.
40 Ruins, Part I, *Lights and Shadows*, II, p. 197.

CHAPTER VI

1   Mrs S.C. Hall, *Sketches of Irish Character*, How & Parsons, London, 1854, Introduction p. x.

2   William Chambers, *Story of a Long and Busy Life*, W. & R. Chambers, Edinburgh and London, 1882, pp. 74–5.
3   *Chambers's Journal*, Vol. IX, No. 435, Sat. 30 May 1840, p. 147.
4   *Edinburgh Review*, Vol. XXXVIII, No. 75, Feb. 1823, p. 105.
5   *Ibid.*, Vol. XII, No. 24, Jul. 1808, p. 410.
6   *Dublin University Magazine*, Vol. XXVI, No. 156, Dec. 1845, pp. 737–52 (p. 747).
7   Mrs Beatrice Grant, *The History of An Irish Family*, Dunbar, Miller, Haddington, Edinburgh, 1822, p. 91.
8   *Eclectic Review*, Vol. VII, Part I, June 1811, p. 537.
9   *Literary Gazette*, No. 703, July 10, 1830, p. 442.
10 *Christian Examiner*, Vol. XI, No. 67, Jan. 1831, pp. 68–9.
11 *Athenaeum*, No. 143, 23 July 1830, p. 457.
12 Mrs S.C. Hall, 'It's Only A Drop', *Stories of the Irish Peasantry*, Chambers, Edinburgh, 1840, p. 39.
13 *Ibid.*'Time Enough', *Stories of the Irish Peasantry*, p. 25.
14 *Ibid.*'Going to Service', *Stories of the Irish Peasantry*, p. 247.
15 William Carleton, 'The Station', *Traits and Stories of the Irish Peasantry*, 2 vols, Tegg, London, 1830, II, p. 234.
16 Mrs S.C. Hall, 'Too Early Wed', *Stories of the Irish Peasantry*, p. 9.
17 *Ibid.*, p. 11.
18 'Sure It Was Always So', *Stories of the Irish Peasantry*, p. 114.
19 'The Landlord at Home', *Stories of the Irish Peasantry*, pp. 92–3.
20 'Family Union', *Stories of the Irish Peasantry*, p. 239.
21 William Carleton, *Parra Sastha; Or The History of Paddy-Go-Easy and His Wife Nancy*, Duffy, Dublin, 1845, pp. 36–7.
22 Mrs. S.C. Hall, 'It's Only the Bite and the Sup', *Stories of the Irish Peasantry*, p. 117.
23 *Ibid.*, p. 122.
24 'The Follower of the Family', *Stories of the Irish Peasantry*, p. 139.
25 *Ibid.*, *Stories of the Irish Peasantry*, p. 152.
26 'The Landlord Abroad', *Stories of the Irish Peasantry*, p. 68.
27 'The Landlord at Home', *Stories of the Irish Peasantry*, p. 82.
28 *Ibid.*, p. 94.
29 *Dublin University Magazine*, Vol. XIV, Oct. 1839, No. LXXXII, pp. 477–79 (p. 479).
30 *Chambers's Journal*, No. 435, 30 May 1840, p. 147.
31 Mrs S.C. Hall, Introduction, *Stories of the Irish Peasantry*, 1850.
32 *Citizen, or, Dublin Magazine*, Vol. IV, No. XXI, pp. 130–33 (p. 130).
33 Mr S.C. Hall, *Retrospect of a Long Life from 1815 to 1883*, 2 vols, Bentley, London, 1883, II, p. 428.
34 *Dublin University Magazine*, 'Our Portrait Gallery', Vol. XVI, No. 92, Aug. 1840, pp. 146–49.

35 Richard Boyse, *Diary*, p. 3. (MSS National Library of Ireland).
36 Mrs S.C. Hall, *Marian, or, A Young Maid's Fortunes*, 2 vols, Collvan, London, 1840, I, pp. 196–97 f.n.
37 *Ibid.*, p. 195.
38 *Citizen*, Vol. I, April, 1840, p. 444.
39 *Chambers' Journal*, No. 435, 30 May 1840, p. 147.

CHAPTER VII
Part I
1 Mr and Mrs S.C. Hall, *Halls' Ireland; its Scenery, Character etc.*, 3 vols, How & Parsons, London, 1843, II, p. 193. (Hereinafter referred to s *Halls' Ireland*).
2 Author's Advertisement, I, p. IV, *Halls' Ireland*.
3 *Literary Gazette*, Sat. 17 Oct. 1840, No. 1239, p. 679.
4 Mr & Mrs S.C. Hall, *Halls' Ireland*, I, p. 1.
5 *Ibid.* I, p. 62
6 *Ibid.* III, p. 487, f.n.
7 *Ibid.* III, p. 487, f.n.
8 *Ibid.* I, p. 129.
9 *Ibid.* II, p. 231.
10 *Ibid.* I, p. 184.
11 *Ibid.* I, p. 212.
12 *Ibid.* I, p. 227.
13 *Ibid.* II, p. 85.
14 *Ibid.* II, pp. 391–92.
15 *Ibid.* III, p. 124.
16 *Ibid.* I, pp. 387–88
17 *Ibid.* III, pp. 281–82.
18 Mr S.C. Hall, *Retrospect of a Long Life from 1815 to 1883*, 2 vols, Bentley, London, 1889, II, p. 100, f.n.
19 *Ibid.*, pp. 426–27.
20 Marilyn Butler, *Maria Edgeworth; A Literary Biography*, Letter from Maria Edgeworth to Fanny Wilson, 23rd July, 1829, quoted in *Maria Edgeworth, A Literary Biography*, Clarendon Press, Oxford, 1972, p. 456.
21 Mr & Mrs S.C. Hall, *Halls' Ireland*, p. 183, f.n.
22 *Ibid.* II, p. 193, f.n.
23 *Ibid.* III, p. 392.
24 *Ibid.* III, p. 393, f.n.
25 William Gilpin, *Three Essays On Picturesque Beauty: On Picturesque Travel; and On Sketching Landscape*, 3rd Edition, Cadell & Davies, 1808, p. 77.
26 Mr & Mrs S.C. Hall, *Halls' Ireland*, p. 135.
27 *Ibid.* II, p. 281.

28 *Ibid.* III, p. 269.
29 *Ibid.* I, p. 46.
30 *Ibid.* III, p. 220
31 *Ibid.* III, p. 219.
32 *Ibid.* II, p. 446.

Part II
1   Mr & Mrs S.C. Hall, *Ireland; its Scenery, Character etc.*, II, p. 127.
2   *Ibid.* II, pp. 3789, f.n. (Hereinafter referred to as *Halls' Ireland*).
3   *Ibid.* II, p. 350.
4   *Ibid.* III, p. 290–91.
5   *Ibid.* III, p. 294.
6   *Ibid.* III, p. 291.
7   *Ibid.* III, p. 48.
8   *Ibid.* I, p. 167.
9   *Ibid.* II, p. 315.
10 *Ibid.* II, p. 315.
11 *Ibid.* III, p. 237.
12 Mrs S.C. Hall, Introduction, *Sketches of Irish Character*, Fifth Edition, How & Parsons, London, 1854, p.xii.
13 *Athenaeum*, No. 3204, 23 March 1889, pp. 375–6.
14 *Ibid.* No. 3206, 6th April 1889, p. 440.
15 Henry Vizetelly, *Glances Back Through Seventy Years; Autobiographical and Other Reminiscences*, Kegan Paul, Trübner & Co. Ltd., London, 1893, p. 305.
16 Mrs S.C. Hall, 'The Mosspits', The *Amulet*, 1832, p. 99.
17 Mr & Mrs S.C. Hall, *Halls' Ireland*, II, p. 413.
18 *Ibid.* II, p. 127.
19 *Fraser's Magazine for Town and Country*, Vol. XXIII, No. 113, Jan. 1841, pp. 90–4.
20 *The Literary Gazette*, No. 1243, 14 Nov. 1840, p. 729.
21 *Dublin University Magazine*, Vol. XVI, No. 96, Dec. 1840, pp. 730–31.
22 *Atlas*, Vol. XV, No. 775, 20 Mar. 1841, p. 191.
23 *The Sunday Times*, 8 Jan. 1843, No. 1,055. p. 2.
24 *Citizen*, 'Mock-Irish Works', Vol. III, No. 18, Mar. 1841, pp. 190–94 (p. 192).
25 *Nation*, 18 Nov. 1843, p. 106.
26 *Dublin Monitor*, Vol. III, No. 385, 22 Apr. 1841, (Lady Chatterton's *Sketches*).
27 Mr & Mrs S.C. Hall, *Halls' Ireland*, II, 371.
28 *Dublin Review*, Vol. XIV, No. 28, Art. I, May 1843, p. 445.
29 Mr & Mrs S.C. Hall, *Ireland; its Scenery, Character, etc.*, 3 vols, George Virtue, London, 1853, III, p. 359.

CHAPTER VIII

1. *Dublin University Magazine*, Vol. XII, No. 68, Aug. 1836, pp. 218–224.
2. Mrs S.C. Hall, *The Whiteboy; A Story of Ireland in 1822*, 2 vols, Chapman & Hall, London, 1845.
3. The *Irish Farmer's and Gardener's Magazine*, Vol. I, No. 1, Nov. 1833, p. 3.
4. *Ibid.*, p. 118.
5. Mary Leadbeater, *Cottage Dialogues among the Irish Peasantry*, Part II, Johnson, London, 1813, p. 108.
6. The *Citizen, Or Dublin Monthly Magazine*, Vol. IV, No. 21, July 1841, p. 133.
7. Mary Leadbeater, *Cottage Dialogues*, Part I, Johnson, London, 1811, p. 216.
8. *Ibid.*, Part II, p. 105.
9. Martin Doyle, *Hints to the Small Holders and Peasantry of Ireland on Road Making and on Ventilation, etc.*, Curry, Dublin, 1830, p. 48.
10. Martin Doyle, *Address to the Landlords of Ireland on Subjects connected with the Melioration of the Lower Classes*, Curry, Dublin, 1831, pp. 30–31.
11. *Irish Farmer's and Gardener's Magazine*, Vol. V, No. 52, Feb. 1838, p. 57.
12. Barbara Hayley, *Carleton's Traits and Stories and the Nineteenth Century Anglo-Irish Tradition*, Colin Smythe, Gerrards Cross, 1983, p. 214.
13. William Carleton, 'The Geography of an Irish Oath', *Traits and Stories of the Irish Peasantry*, 2 vols, Tegg, London, 1830, Vol. II, p. 63.
14. Mrs S.C. Hall, 'Kelly the Piper', First series, Westley Davies, London, 1829, *Sketches of Irish Character*, I, p. 330.
15. 'The Bannow Postman', *Sketches of Irish Character'*, First Series, Westley & Davis, London, 1831, p. 158.
16. 'Mabel O'Neill's Curse', *Sketches of Irish Character*, Second Series, Westley & Davis, London, 1831, p. 316.
17. 'Kate Connor', *Sketches*, Second Series, p. 43.
18. 'Annie Leslie', *Sketches*, Second Series, p. 84.
19. *Irish Farmer's and Gardener's Gazette*, Vol. I, No. 6, Apr. 1834, p. 312.
20. Mrs S.C. Hall, 'The Bannow Postman', *Sketches*, First Series, I, p. 159.
21. *Farmer's Gazette and Journal of Practical Horticulture*, Vol. II, No. 36, 6 Jan. 1844, p. 284.
22. *Irish Farmer's and Gardener's Magazine*, Vol. I, No. 3, Jan. 1834, p. 118.
23. Mrs S.C. Hall, *The Whiteboy*, II, pp. 288–289.
24. *Ibid.* I, p. 106.
25. *Ibid.* I, p. 72
26. *Ibid.* I, p. 32.
27. *Ibid.* I, pp. 238–239.
28. *Ibid.* I, pp. 64–65.
29. *Ibid.* I, p. 250.
30. *Ibid.* I, p. 126.

31 Barbara Hayley, 'Religion and Society in Nineteenth Century Irish Fiction', edited by Robert Welch, Colin Smythe, Gerrards Cross, 1990.
32 Mrs S.C. Hall, *The Whiteboy*, II, pp. 295–296.
33 *Ibid*. I, p. 57.
34 *Ibid*. I, p. 146.
35 *Ibid*. I, p. 144.
36 'Too Early Wed', *Stories of the Irish Peasantry*, p. 13.
37 *The Whiteboy*, I, p. 39.
38 *Ibid*. I, p. 178.
39 *Ibid*. II, p. 8.
40 *Ibid*. II, pp. 222–223.
41 *Ibid*. I, p. 119.
42 *Ibid*. I, p. 317.
43 *Ibid*. II, p. 32.
44 *Ibid*. I, p. 163.
45 *Ibid*. II, p. 162.
46 *Ibid*. I, p. 232.
47 'Too Early Wed', *Stories of the Irish Peasantry*, p. 13.
48 Lady Morgan (Sydney Owenson), *The Wild Irish Girl*, 3 vols, reprint of 1806 edition, Garland Publishing Inc., New York and London, 1979, III, pp. 258–259.
49 Charles Maturin, *The Wild Irish Boy*, 3 vols, reprint of 1808 edition, Garland Publishing Inc., New York and London, 1979, III, pp. 138–139.
50 *Athenaeum*, No. 50, Oct. 1828, pp. 788–790.
51 Mrs S.C. Hall, *The Whiteboy*, I, p. 33.
52 *Ibid*., II, p. 306.
53 *Ibid*. II, p. 300.
54 *Ibid*. I, p. 27.
55 *Ibid*. I, p. 9.
56 *Ibid*. I, p. 53.
57 *Ibid*. I, p. 21.
58 *Ibid*. I, pp. 306–307.
59 *Tablet*, Vol. VI, No. 275, Aug. 1845, pp. 499–500.
60 *Spectator*, Vol. XVIII, No. 893, Aug. 1845, p. 469.
61 *Pilot*, 18 Aug. 1845.
62 *The Sunday Times*, No. I, 194, 7 Sept. 1845, p. 2.
63 *Literary Gazette*, No. 1,490, Aug. 1845, p. 521.
64 *Athenaeum*, No. 929, 16 Aug. 1845, p. 810.
65 *Gentleman's Magazine*, Vol. XXIV, N.S., Nov. 1845, p. 507.
66 *John Bull*, Vol. XXV, No. 1,288, 16 Aug. 1845, p. 520.
67 *Atlas*, Vol. XX, No. 1,004, 9 Aug. 1845, p. 507.
68 Sir Samuel Ferguson, 'The Didactic Irish Novelists', *Dublin University*

Magazine, Vol. XXVI, No. 156, Dec. 1845, pp. 737–752 (p. 748).
69 *Athenaeum*, No. 552, 26 May, 1838, p. 398 (review of *Lights and Shadows of Irish Life*).

CHAPTER IX

1   William Carleton, *Valentine M'Clutchy; The Irish Agent, Or, The Chronicles of Castle Cumber*, 3 vols, Duffy, Dublin, 1845. (Hereinafter referred to as *Valentine M'Clutchy*).
2   Charles Lever, *The O'Donoghue; A Tale of Ireland Fifty Years Ago*, Curry, Dublin, 1845, p.
3   Charles Lever, *St Patrick's Eve*, Chapman & Hall, London, 1845.
4   Charles Lever, *The O'Donoghue*, p. 124.
5   *Ibid.*, p. 109.
6   William Carleton, *Valentine M'Clutchy*, I, pp. 125–126.
7   Charles Lever, *St Patrick's Eve*, Chapman & Hall, London, 1845, p. 124.
8   *Ibid.* p.52
9   William Carleton,*Valentine M'Clutchy*, III, p. 326.
10 Oliver MacDonagh, *The Nineteenth-Century Novel and Irish Social History*, National University of Ireland, Dublin [n.d.] p. 7.
11 *Atlas for India*, Vol. XX, No. 975, Jan. 1845, p. 42.
12 *Spectator*, Vol. XVIII, No. 863, Jan. 1845, pp. 41–42.
13 Charles Lever, *St Patrick's Eve*, p. 198.
14 *Literary Gazette*, No. 1,472, Apr. 1845, pp. 211–212.
15 *Atlas*, Vol. XX, No. 986, Apr. 1845, p. 217.
16 *The Sunday Times*, No. 1,177, 11 May 1845, p. 2.
17 *Athenaeum*, No. 915, May 1845, p. 460.
18 *Spectator*, Vol. XVIII, No. 874, Mar. 1845, p. 305.
19 *Nation*, 12 Apr. 1845, p. 443.
20 Charles Lever, *The O'Donoghue*, p. 409.
21 *Ibid.*, p. 226.
22 *Nation*, 12 Apr. 1845, p. 443.
23 William Carleton, Preface, *Valentine M'Clutchy*, p. xi.
24 *Ibid*. II, p. 274.
25 *Ibid*. I, p. 26.
26 *Ibid*. I, p. 42.
27 *Athenaeum*, No. 898, Jan. 1845, pp. 38–39.
28 *Tait's Edinburgh Magazine*, Vol. XII, No. 135, Mar. 1845. p. 172.
29 William Carleton, *Valentine M'Clutchy*, III, pp. 297–8.
30 *Ibid.*, p. 226.
31 *Ibid.*, Preface, p. vii.
32 *Tait's Edinburgh Magazine*, Vol. XII, No. 135, Mar. 1845, p. 172.
33 William Carleton, *Valentine M'Clutchy*, III, pp. 26–7.
34 *Ibid*. III, p. 313.

35 *Atlas for India*, Vol. XX, No. 975, 18 Jan. 1845, p. 42.
36 Charles Lever, *Charles O'Malley; The Irish Dragoon*, 2 vols, Curry & Co., Dublin, 1841, I, p. 58.
37 Charles Lever, *The O'Donoghue*, p. 207.
38 *Tablet*, Vol. VI, No. 252, 8 Mar., 1845, p. 147.
39 *Dublin Review*, Vol. I, May 1836, p. 86.
40 Charles Lever, *The O'Donoghue*, p. 238.
41 *Ibid.* p. 401.
42 William Carleton, *Valentine M'Clutchy*, II, p. 283.
43 *Ibid.* II, p. 142.
44 *Ibid.* III, p. 141.
45 *Ibid.* III, p. 147.
46 *Ibid.* III, p. 103.
47 *Ibid.* III, p. 156.
48 *Ibid.* I, p. 49.

CHAPTER X
1 Henry Vizetelly, *Glances Back Through 70 Years, Autobiographical and Other Reminiscences*, 2 vols, Kegan Paul, Trench, Trübner & Co. Ltd., London, 1839, I, pp. 305–6.
2 Thomas Moore, *Memoirs* (ed. Russell), 8 vols, Longman, London, 1856, VII, p. 315.
3 W.P. Frith, *My Autobiography and Reminiscences*, 4 vols, Bentley, London, 1887, II, pp. 346–7.
4 *The Illustrated News of the World*, 1861.
5 Alice King, quoted in W.H. Goss, *The Life and Death of Llewellynn Jewitt, F.S.A.*, Henry Craig, London, 1889. (Hereinafter referred to as *Jewitt*),pp. 365–6.
6 John Forster, *The Life of Charles Dickens*, 3 vols, Chapman & Hall, London, 1873, II, p. 444.
7 S.C. Hall, *Retrospect of a Long Life; 1815 to 1883*, 2 vols, Bentley, London, II, 1889, pp. 67–70. (Hereinafter referred to as *Retrospect*).
8 Peter Ackroyd, *Dickens*, Sinclair-Stevenson, London, 1990, p. 842.
9 S.C. Hall, *Retrospect*, II, p. 156.
10 *Retrospect*, I, p.90.
11 *Punch, or, The London Charivari*, Vol. X, Jan–Jun, 1846, p. 149.
12 Mrs. S.C. Hall, *Midsummer Eve; A Fairy Tale of Love*, Longman, London, 1847.
13 *Athenaeum*, No. 1050, 11 Dec. 1847, p. 1270.
14 *Godey's Magazine and Lady's Book* (Philadelphia, USA), Vol. XLV, Aug. 1852, pp. 134–136.
15 W.H. Goss, *Jewitt*, p. 385.
16 Mrs S. C. Hall, 'Papa's Letter', *The Juvenile Budget*, 1840, p. 254.

17 'The Private Purse' 'Miniature Library of Fiction', Vol. III, Chambers, Edinburgh, 1858, p. 14.
18 Ibid., p. 20.
19 Wives and Husbands, 'Miniature Library of Fiction', Vol. IX, Chambers, Edinburgh, p. 25.
20 W.H. Goss, *Jewitt*, p. 432.
21 Mrs S.C. Hall, *The Private Purse*, pp. 37–38.
22 Ibid., p.40.
23 *The Bloomsbury Dictionary of Women Writers*, Bloomsbury Publishing Ltd., London, 1992, p. 615.
24 *The Dictionary of National Biography*, Vol. VIII, p. 938.
25 Michael Scott, Introduction, *Hall's Ireland; its Scenery, Character, etc.*, 2 vols., Sphere Books, London, 1984, I, p. xiv.
26 W.H. Goss, *Jewitt*, p. 432.
27 *Athenaeum*, No. 1020, 15 May 1847, p. 517.
28 Mrs S.C. Hall, 'The Cry From Ireland', *Art Union*, Apr. 1847, p. 141.
29 Mrs S.C. Hall, Introduction, *The Whiteboy*, 2 vols, Chapman & Hall, London, 1855.
30 Mrs S.C. Hall, *The Fight of Faith*, 2 vols, Chapman & Hall, London, 1869, Dedication.
31 Mrs S.C. Hall, Letter to Francis Bennoch, 1853, MSL, H. 17 be (University of Iowa).
32 Review of *The Fight of Faith*, unattributed, undated, newspaper cutting in Mrs Hickey's Scrapbook, MS 407 (NLI).
33 *Athenaeum*, No. 2153, 30 Jan. 1869, p. 170.
34 Mrs S.C. Hall, 'God Save the Green; A Few Words to the Irish People', Twopenny Illustrated Tracts, S.W. Partridge, London, ca. 1871.
35 *Irish Builder*, Vol. 21, 1 May 1879, pp. 139–9.
36 *The Times*, 1 Feb. 1881, p. 10.
37 *Irish Times*, Vol. XXIII, No. 6599, 2 Feb. 1881., p. 4.
38 *Wexford Independent*, Vol. LII, No. 2715, 5 Feb. 1881, p. 6.
39 *Irish Builder*, Vol. XX, No. 508, 15 Feb. 1881, p. 50.
40 *Athenaeum*, No. 2780, 5 Feb. 1881, p. 200.
41 *Irish Builder*, Vol. XXXI, 1 May 1889, p. 124.
42 Horatio Krans, *Irish Life and Irish Fiction*, Macmillan, New York and London, 1903, p. 116.
43 Stephen J. Brown, S.J., *Ireland in Fiction*, Irish University Press, Shannon, Ireland (Facsimile of Second Edition 1919), pp. 125–6.
44 J.C. Twomey, 'The Bay and Town of Bannow', *Kilkenny Archaeological Transactions*, Vol. I, No. II, 1849–1851, pp. 194–203.
45 James Newcomer, 'Mrs Samuel Carter Hall and *The Whiteboy*' in *Etudes Irlandaises*, vol. VIII, Dec 1983, pp. 113–119.
46 Barry Sloan, *Pioneers of Anglo-Irish Fiction*, 1800–1850, Colin Smythe,

1986, Barnes & Noble Books, Totowa, N.J. p. 202.
47 Barry Sloan, 'Mrs Hall's Ireland', *Eire/Ireland*, Irish Association Cultural Institute Quarterly, No. XIX, i, 1984, pp. 18–36.
48 Gregory Schirmer, 'Tales From Big House and Cabin: The Nineteenth Century' *The Irish Short Story: A Critical History*, ed. James F. Kilroy, Twayne Publishers, Boston, 1984, pp. 21–44 (p. 22).
49 Robert Lee Wolff, 'Introduction', *Sketches of Irish Life and Character*, p. xvii.
50 *Tait's Edinburgh Magazine*, Vol. II, No. XI, Feb. 1833, p. 572.

Index

Page references in **bold** indicate complete chapters; AMH refers to Anna Maria Hall and SCH refers to Samuel Carter Hall. Works by anonymous authors are to be found under their titles.

Ackermann, Richard (German lithographer) 27
Ackroyd, Peter, *Dickens* 201
Act of Union 88, 185, 187-88
Adburgham, Alison, *Silver Fork Society* 76
Adelphi Theatre (London) 9, 80
Aesop, *Fables* 45
Agricultural Society 153
Ainsworth, Harrison, *Jack Sheppard* 46
Albert, Prince Consort 201
Amulet 28, 30; AMH writes for 26, 27, 57, 139-40; collapse 19, 25
Annuals 27-30, 31, 32, 57, 64
Art Journal (later *Art Union*) 19, 32, 201
Art Union (previously *Art Journal*) 124, 203, 206; SCH edits 11, 19, 201, 202
Athenaeum 66, 76, 98, 168; attitude to morality 42; letters to 138-39; public taste 46; reviews *The Fight of Faith* 209-10; reviews *Lights and Shadows of Irish Life* 81; reviews *Marian* 10; reviews *Midsummer Eve* 202; reviews Miss Mitford 63, 64; reviews *St Patrick's Eve* 183; reviews *Sketches of Irish Character* 8, 14; reviews *Valentine M'Clutchy* 188; reviews *The Whiteboy* 174; tribute to AMH 211; wearies of Irish subjects 206
Atlas 11, 29-30, 57, 141, 202
Atlas for India 182, 191
Austen, Jane 14, 45, 46; *Mansfield Park* 38; *Northanger Abbey* 43

Banim, John and Michael ('The O'Hara Brothers') 3, 12, 49, 206
Barbauld, Mrs 14

Barrow, Sir John, *Account of a Journey in Africa* 1
Battle of the Boyne 129-30, 208, 209
Bayley, Frederick, *Four Years in the West Indies* 2
Baynes, Messrs (publishers of religious works) 27
Beckett, J.C., *Bloomsbury Dictionary of Women Writers* 205
Bell, Mrs Martin, *Julia Howard* 45
Bentley, Richard 75-76
Bentley's Miscellany 42
Betham, Sir William 114
Bible, The 35
Blackwood's Edinburgh Magazine 2, 27; attitude to moral teaching 41, 43, 46; *Noctes Ambrosianae* 77; reviews Miss Mitford 62, 63
Blessington, Marguerite, Countess of 28, 75, 201
Bligh, Captain William 1
Book of Beauty (Annual) 27, 28
Book of Gems (Annual) 28
Book of Royalty 32
Boons and Blessings (temperance tract) 21
Boyse, Samuel 87
Boyse, Thomas 6-7
Bremer, Frederika 42, 46
British Library 98
Brittaine, George 12, 157; *Hyacinth O'Gara* 39; *Irish Priests and English Landlords* 39
Brittania 18
Brompton Hospital for Consumption (now Brompton Hospital) 203
Brontë, Anne 12

251

Bronté, Charlotte 11, 12
Bronté, Emily 12
Brooke, Henry, *The Fool of Quality* 37
Brooke, W.H. 113
Brown, Stephen J., *Ireland in Fiction* 211
Browning, Elizabeth Barrett 28
Browning, Robert 28
Bruce, James (explorer) 1
Bulwer-Lytton, Edward 75, 88; *Paul Clifford* 8; *Pelham* 30
Bunyan, John 48
Burckhardt, Johann Ludwig, *Arabic Proverbs* 2
Burke, Edmund, *Philosophic Enquiry into the Origin of Our Ideas of the Sublime and the Beautiful* 2
Burke's Peerage 76
Burney, Fanny 14
Bury, Lady Charlotte 3, 75
Butler, Father (Wexford historian) 7, 84
Butler, Marilyn 122
Butler, W.A. 109-10
Butt, Isaac 34
Byron, Lord (George Gordon) 28

Cahalan, James M., *Great Hatred, Little Room* 211
Campbell, Thomas; 'Hohenlinden' 88; 'Lord Ullin's Daughter' 88; 'Ye Mariners of England' 88
Carleton, William 3, 12, 49, 198, 212; attitude to priests 189-91, 196-97; didacticism 103-04, 197-98, 206; in Irish periodicals 33, 34; landlord/tenant relationship 150-51; schoolmasters 166; sectarianism 190, 191; use of dialect 70; violence in 192, 194-97; *Art Maguire* 48, 175, 197; *The Black Prophet* 162, 206; 'The Geography of an Irish Oath' 150-51; 'Larry M'Farland's Wake' 150; *Parra Sastha* 97, 103-04, 175, 197; *The Poor Scholar* 106; *Roddy the Rover* 175, 197; *St Patrick's Eve* 197; 'Shane Fadh's Wedding' 150; *The Squanders of Castle Squander* 206; 'The Station' 100; *Tales for the Irish People* 48; *Traits and Stories of the Irish Peasantry* 12, 100, 150, 213; 'Tubber Derg' 181; *Valentine M'Clutchy* 11, 177, (didacticism) 197-98, (landlords and agents) 179-82, 183, 184, 187-88, 195, 196, (politics) 186-88, (religion) 188-91, (violence) 194-97
Carr, George (step-grandfather of AMH) 5-6, 18, 110
Carr, Mrs George (grandmother of AMH; second wife of GC) 4, 5-6, 27, 84
Carr, Sir John, *Stranger in Ireland* 1
Carr, Reverend George 58, 86
Catholic Emancipation 86
Chamber's Journal 95, 112; AMH writes for 10, 31, 108, 109, 202
Chambers, Robert (publisher) 95
Chambers, William (publisher), *Story of a Long and Busy Life* 95
Chatterton, Lady, *Irish Sketches* 142
Christian Examiner 39, 69, 98
Citizen 109, 111, 142
Clare, John 25
Cobbett, William 96
Colburn, Henry (publisher) 1, 75-76, 77, 88
Coleridge, Samuel Taylor 25, 28, 200
Committee of Inquiry into Deep Sea Fisheries 114
Cook, Captain James 1
Cowper, William, *The Task* 36
Crabbe, George: *The Borough* 36; *The Parish Register* 36; *Tales of the Hall* 36; *Tales in Verse* 36
Croker, John Wilson 114
Croker, Thomas Crofton 12, 78, 114, 120; *Fairy Legends and Traditions of the South of Ireland* 69, 138
Croly, Reverend George, *Tales of the Great St Bernard* 39
Cromwell, Oliver 129
Crowe, Eyre Evans 12, 145
Cunningham, Allan 8; 'Biographical and Critical History of the Literature of the Last Fifty Years' 63
Curtis, L. Perry, Jnr, *Apes and Angels, the Irishman in Victorian Caricature* 54

Davey, Hannah (servant to the Halls) 22
de Latre, G. (painter) 200
Derrick, Mr (local historian) 113
Derry, Siege of 129, 130
Dickens, Charles 12, 28, 44, 46-47, 139, 200-201; *Nicholas Nickleby* 110-11

Index

Dickens, Mrs Catherine 201
Dictionary of National Biography 205
didacticism: an age of 14; attack on 48; critical confusions about 43; fiction for Irish peasants 97-98; in literature 35-49; political 188
Disraeli, Benjamin 44, 75
D'Orsay, Alfred, Count 28
Doyle, Martin *see* Hickey, Reverend William
Doyne, Charles 153
Dublin Literary Gazette 32, 33, 62
Dublin Monitor 142-43
Dublin Monthly Magazine 39
Dublin Penny Journal 30, 34
Dublin Review 34, 192-93
Dublin University Magazine 33-34, 97; attitude to moral teaching 44, 48, 49; portrait of AMH 109, 111; reviews *Ireland: its Scenery, etc.* 141; reviews *Lights and Shadows of Irish Life* 82, 145; reviews *Stories of the Irish Peasantry* 108; reviews *The Whiteboy* 174, 176, 197
Duff and Company, (publishers) 77
Dunsany, Lady 22

Easter Gift, The (Annual) 28
Echoism (Anon) 43
Eclectic Review 4, 8, 38, 41, 46, 62
Edgeworth, Maria 12, 25, 38, 46, 49, 145; and AMH 11, 13-14, 121-22, 174, 211; humanizes the Irish 9; indifference to religion 39, 41; moral teaching 39, 41-43, 45; as social commentator 41-42; *The Absentee* 41, 57; *Castle Rackrent* 41, 121, 212; *Ennui* 13, 57; *Manoeuvring* 45; *Ormond* 57; *Patronage* 45; Preface to *Cottage Dialogues among the Irish Peasantry* 97
Edinburgh Literary Gazette 13, 41
Edinburgh Literary Journal 8-9, 13
Edinburgh Review 1, 12, 37, 96-97; attitude to moral teaching 40, 41, 42
Elizabeth, Charlotte 157
Evelyn, John, *Diary* 76
Examiner 75

Fair Carew, The (Anon) 44
Fairholt, Mr 124
famine 160-62, 206-08
Faucit, Helen 200

Faxon, Frederick Winthrop 28-29
Fenianism 207, 210
Ferguson, Sir Samuel 48, 49, 97, 176; 'Didactic Irish Novelists' 174-75
Fielding, Anna Maria *see* Hall, Anna Maria
Fielding, Henry, *Tom Jones* 36-37
Fielding, Mrs (mother of AMH) 5, 6, 21, 23, 27
Finden's Tableaux (Annual) 28, 30
Fitzgerald, Mr (local historian) 113
Flanagan, Thomas, *The Irish Novelists* 211
Fletcher, Giles 36
Florestan (Anon) 40
Forget-me-Not (Annual) 27
Forster, John, *The Life of Charles Dickens* 200
Foscolo, Ugo (exiled Italian poet) 17, 18
Fowler, Robert 150
Fraser's Magazine for Town and Country 2, 12; attitude to moral teaching 42, 44, 45; reviews *Finden's Tableaux* 30; reviews *Ireland: its Scenery, etc.* 10-11, 141; reviews *Sketches of Irish Character* 8
Friendship's Offering (Annual) 27, 29, 64
Frith, W.P. 199-200

Galt, John, *Sir Andrew Wyllie* 40
Garland Publishing Company (New York and London) 212
Gaskell, Elizabeth, *Mary Barton* 44, 47
Gentleman's Magazine 11, 63; attitude to moral teaching 39, 41, 49; reviews *The Whiteboy* 14, 174
George IV, King of Great Britain 26
Gift Books *see* Annuals
Gilpin, William 124
Giraldus Cambrensis 113
Godey's Magazine and Lady's Book 203
Godwin, William 3
Goethe, Johann Wolfgang von 48; *Werther*, 40
Goldsmith, Oliver, *The Vicar of Wakefield* 37
Gore, Mrs Catherine 14, 75; *Stokehill Place, or the Man of Business* 41
Goss, W.H., *The Life and Death of Llewellynn Jewitt, F.S.A.* 200, 203-04, 205

Governesses' Benevolent Institution Asylum 203
Gower, John 36
Grace, Sheffield, 'Memoirs of the Grace Family' 114
Grant, Mrs Beatrice, *The History of an Irish Family* 97
Graves, Dean 157, 170, 175
Griffin, Gerald 3, 12, 49, 159, 206
Grogan Morgan family 7, 102, 107, 120
Grogan Morgan, Lizzy 23
Grogan Morgan, Mrs 92
Grose, Francis 114

Hall, Anna Maria (née Fielding; wife of SCH) : birth 1, 5, 135, 210; childhood in County Wexford 1, 4-5, 6, 55, (memories) 60, 63; leaves Ireland 88, 110, 135; schooling in England 110; marriage 17, 21, 77, 110, (devoted wife) 19, 27, 203, (domestic duties) 23-24, (entrée to literary world) 19, (happy) 135, 140, (worries about SCH) 88; health 26-27; children 21-23, 88; at Upper Charlotte Street 33; in Chelsea 27; at the Rosery 199-200, 202, 203; social life 199-202; in Dickens's house 200-201; at Bannow Lodge 200; at Campden Hill 22; friends 87, 102, 107, 120-21, 139, 199-200, 206; trips to Ireland 9, 31, 113, 115, 120-21, 133, 206; portrait 200; photograph engraving 200; pension from Civil List 19; golden wedding gift 19; death 18, 22, 200, 210; obituaries 210-11; tributes 200, 210:
ATTITUDES AND THEMES: absenteeism 11, 56-57, 127, 128, 153; adventurous tourist 117-19; agents 56-57, 91, 127, 128, 152-5 158-9; ambivalence about nationality 135; beggars 70-71, 89-90, 132, 140, 141, 160; bigotry 12, 80, 82, 109, 129, 208-09, 212; children 52, 56, 61, 99, 100-101; class-consciousness 52-54, 56-57, 82, 159; colonial **51-73**, 79, 82-84, 212; compassion for poor 26; conservatism 102-03, 156; cottages 7, 8, 52, 56, 79, 91, 132, 156; to countrymen and women 72; devout Christian 22, 23, 66, 198; to divorce 203, 204; endurance and stamina 115; English superiority 72, 79, 104-05, 156, 169-70, 175; enthusiasms 111, 126, 202-03; false pride 83-84, 100, 135-36; famine 160-62, 208; hard working 199, 202; to her 'ruined country'; hospitality 105, 132; to illegitimacy 59-60, 66; independence 56, 84; indolence 55, 58, 83, 133, 135, 156, (denies comforts) 56, 57; intemperance 58, 60, 78-79, 86, 128-29, 210, (peasant) 51, 83, 133; Irish failings 58-59, 79, 80-81, 83-84, 98-100, 128, (correcting) 10, 95-96, 108-09, 168, 194; Irish virtues 57, 59, 80, 83-84, 105-06, 131; Irish women 59-60, 83, 135; landlord/tenant relationship 7, 56-57, 91-92, 105, 108-09, 151, (central to writing) 127-28, 145-47; love of Ireland and Irish 208, 211; marriage 68-69, 101-02, 135, 204-05; nationalism 33, 34, 169; old/new landlords 151-52, 156; optimism 207-08, 210; philanthropy 27, 203, 207; politics 11, 167-69, 171, 185-86; popularity 10, 109, 110; poverty 54, 56, 82, 88-90, 132-33, 135; priests 4-5, 67-69, 84-85, 127, 189; procrastination 54, 58, 79; racism 52; realist 139-40; religion 67, 143, 159, 169-71; religious tolerance 12, 69, 109-10, 129, 156, 188, 191, (family tradition) 84-85, (personal views) 198, 209, (priests and) 126-27; schools/schoolmasters 110-11, 159, 161, 165-66; secret societies 165-66; servants 7, 10, 24-25, 60-61, 106, 136; spiritualism 22; superstition 11, 24, 67, 69-70, 83, 102, 138; temperance 78, 86; tourism 81-82, 87, 113, 122-23, 131, 144; violence 131, (agrarian) 28, 64, 88-89, 162-63, 193, (appetite for) 9, 61-62, 64, 65-66, (Irish propensity for) 61, (political) 64-65, 163-64, 193, (rebellion) 4-5, 60, 63, 64, 67, 166, (revenge/sexual) 66, (secret societies) 164-66; wifely independence 203-04, 205, 214; women's rights 203, 205:
WRITING: an apprenticeship for *The Whiteboy* 145; articles 206-07; assessment of 109; autobiographical

Index

5, 19-20; children's stories/books vii, 8, 23, 202, 203, 210; collaboration with SCH 3, 68, 111-12, 131, 214, (in obituaries) 210, (published) vii, 10, (success) 143, 203; collections published vii, 10, 20, 76-77, 92, 95, 167; 'cottage girl portraits' 52-53; criticism of 142-43, 176; didacticism 14-15, 44, 49, 62, 92, 158-59, (assessments) 213, (central to work) 35, 49, (criticism of) 80, (in *Stories of the Irish Peasantry*) 96, 98-99, 102, 104, 108-09, (of the colonist) 55-56, (political) 168-69, 174-75, 176; drama 9, 32, 77-78, 80; 'feminine delicacy' 12, 34, 198; folk tales 69, 114; guide books vii, 3, 10, 119; guilty of plagiarism 8; heroines 59-60, 70, 81, 82-83, 98-99, 161; influence of Maria Edgeworth 121-22; influence of SCH 138-40; interpreter of Irish life and character vii, viii, 9, 10-11, 33, 82-83, (assessments) 211, 214, (authenticity) 7, (personal views) 136, (success) 110, 114; Irish scenery 3-4; 'Irishness' 8, 14, 15, 34, 110, 213; journalism vii, 31, 114, 176, 210, (editorial) 26, 30, 32, 202; journey into adulthood 134; language/dialect 12, 81, (assessments) 211, 212, 213, (Cockney) 81, (lurid) 208-09, (peasant) 59, 70-72, 83, 107, (priests) 85-86, (vigorous) 99, 137, (violent) 65; letters 32-33, 121, 204-05, 209; literary début 25; literary skills 99, 111; a literary teacher 104; market for 30, 32, 33, 34; a national teacher 174; novella 203; novels (didactic) 197, 198, ('English') vii, 76, 145, (last) 208, 210, (political) 173, 176, 186, 207; of literary significance 205-06; popularity 30, 213; reputation 32, 34, 109, 142; reviews and assessments 211-12; success 32, 202, 214; women writers 14:

WORKS: 'Annie Leslie' 57, 153; 'The Bannow Postman' 52, 58, 69, 98, 152, 154; 'Beggars' 88; 'Black Dennis' 20, 65; *The Book of the Thames* 202; *The Buccaneer* 31, 76, 145; 'Building a House with a Tea-cup' 21; 'Captain Andy' 59, 67; *Chronicles of a Schoolroom* 8; 'The Cry from Ireland' 206; 'Debt and Danger' 96, 106; 'Dermot O'Dwyer' 86; 'Digging a Grave with a Wine Glass' 21; 'The Dispensation' 84, 85; 'The Drowned Fisherman' 30; 'The Fairy of Forth' 58, 69; 'Family Union' 103; 'Father Mike' 58, 59, 64-65; 'The Follower of the Family' 105, 106; *The Forlorn Hope* 203; *The French Refugee* (play) 32; 'The Gipsey Girl' 26; *God Save the Green* 210; 'Going to Service' 99-100, 104; 'Good Spirits and Bad' 58; *Harry O'Reardon* 106; 'Hospitality' 52, 61, 152; 'Independence' 56, 98; *Ireland: its Scenery, etc.* 10-11, 112, **113-44**, 214, (condensed version) 205, (descriptions of scenery) 123-24, (essays) 140, (famine) 208, (figures) 124-26, (German translation) 143, (influence of Rosa) 3, (inns and hotels) 116-17, 122-23, (new edition) 143-44, (published) vii, (reviews) 141-43, 203; 'It's My Luck' 93, 98; 'It's Only A Bit Of A Stretch' 104-05; 'It's Only the Bite and the Sup' 105; 'It's Only a Drop' 98; 'Jack the Shrimp' 66; *The Juvenile Budget* 23, 24, 204; 'Kate Connor' 32, 33, 56-57, 153, 154, 181; 'Kelly the Piper' 58, 107, 140, 151, 152; 'The Landlord abroad' 104, 107; 'The Landlord at Home' 102-03, 107; 'Larry Moore' 55; 'The Last in the Lease' 89; 'The Last of the Line' 59, 61, 153; *Lights and Shadows of Irish Life* 9, 31, **75-93**, 104, 128, 134, (facsimile edition) 212, (published) vii, 9, 31, 95, (reviews) 145, (secret societies) 131, (servants) 106, 136; 'Lilly O'Brien' 52-54, 70, 72; 'Mabel O'Neill's Curse' 32, 53, 60, 66, 152; *Marian; or, A Young Maid's Fortunes* 10, 31, 110-11, 145; 'Mary Clavery's Story' 20; 'Mary Ryan's Daughter' 59-60; 'Master Ben' 19, 20, 26; *Midsummer Eve: A Fairy Tale of Love* vii, 11, 176, 202; 'The Mosspits' 28, 139-40; 'Moyna Brady' 93; 'The Murmurer Instructed' 26; 'The Old Clock' 30; 'Old Frank' 60, 69; 'The Old Governess' 203; *The Outlaw* 31, 76, 145; 'Papa's Letter' 204; 'Peter

the Prophet' 52, 58; *The Private Purse* 204, 205; 'The Rapparee' 3, 8, 65-66, 70; 'The Savoyards' 26; 'Servants' 136; *Sketches of Irish Character* 58, 104, 128, 136; (Series 1) 3, 5, 31, 60, 61-62, (S1 facsimile edition) 212, (S1 published) vii, 4, 51, (S1 reviews) 7, 9, 13; (Series 2) 65-66, 121-22, (S2 published) vii, 3, 4, 51, 62, (S2 reviews) 8, 13-14, 57, 62; (Series 3) 5, 6-7; (Series 5) 5, 20, 24, 26, 84, 95; (colonial/class attitudes) 82-84; (illustrations by Maclise) 214; (landscape) 107; (tragic heroines) 59-60; 'Sketches on Irish Highways' 87; 'Star' 26; *Stories of the Irish Peasantry* vii, 10, **95-112**, 162, 167, 203, 212; 'Sure It Was Always So' 102, 105; 'Take it Easy' 69; *Tales of a Woman's Trials* 8-9, 32; *The Fight of Faith* 186, 209, 212, (published) vii, 11-12, 176, 202, 208; *The Groves of Blarney* 9, 32, 77-83, 98-99; *The Whiteboy* 3, **145-76**, 156, (agents) 152-55, 158-59, 178, 180, 190, (assessments) 211-12, (didacticism) 44, 158, 173, 174-75, 197, 198, (English superiority) 160, 169-70, 175, (facsimile edition) 212, (landlords and tenants) 92, 145, 151, 158, 176, 181, (new edition) 207-08, (peasants) 166-67, 170, 172, (politics) 169, 173-74, 176, 185, 186, (poverty) 161-62, (priests) 157-58, (published) vii, 14, 177, 202, (religious tolerance) 158-59, 169-70, 176, 188, (reviews) 10, 11, 48, 173-76, 197, (violence) 162-67, 176; 'Time Enough' 99; 'Too Early Wed' 101-02, 105, 167; *Uncle Horace* 31, 76, 145; *A Week at Killarney* 11; 'We'll See About It' 54, 57, 98; 'The Wise Thought' 52, 53; *Wives and Husbands* 204; 'The Wooing and the Wedding' 72; 'Young Rebel' 26
Hall, Captain Basil, *Voyage to Loo-Choo* 2
Hall, Colonel (father of SCH) 17, 22
Hall, Joseph 36
Hall, Maria Louise (daughter of AMH) 22
Hall, Mrs (mother of SCH) 17
Hall, Mrs Samuel Carter *see* Hall, Anna Maria
Hall, Samuel Carter: birth 17, 135; experience of rebellion 17; leaves Ireland 17; marriage 17, 18, 21, 77, 135, 140; children 21-23, 88; at the Rosery 199-200, 202, 203; social gatherings 199-202; in Dickens's house 200-201; at Bannow Lodge 200; at Campden Hill 22; friends 87, 102, 107, 120-21, 139, 199-200, 206; trips to Ireland 9, 31, 113, 115, 120-21, 133, 206; monument to father 17, 22; pension/retirement 19, 201; death 138, 211; obituaries 139, 211:
ATTITUDES: adventurous tourist 117-19; charitable activities 203; devout Christian 22; an 'Englishman' 135; enthusiasms 126; to famine 208; optimism 208; religious/political tolerance 129-30; to servants 25; spiritualism 211:
CAREER: literary/parliamentary secretary 17, 18; journalism vii, 114, (editorial) 9, 11, 18-19, 20, 28; founds *Amulet* 19; *Amulet* collapses 19, 25; elected Fellow of Society of Antiquaries 78, 120; edits *Art Union* 11, 19, 32, 201, 202; unsuccessful ventures 18-19, 31, 202:
WRITING AND REPUTATION: articles 18; charged with hypocrisy 202; collaboration with AMH 3, 68, 111-12, 131, 214, (in obituaries) 210, (published) vii, 10, (success) 143, 203; essays 32; good editor 25; hard worker 199; history of France 32; influence on AMH 138-40; nicknamed Mr Pecksniff 202; on popularity of AMH 109; poetry 32; pomposity and sententiousness 139:
WORKS: 'Death Cry of Alcestis' 28; 'Death of Eucles' 28; 'The Dying Girl to Her Mother' 28; *Ireland: its Scenery, etc.* 10-11, 112, **113-44**, 214, (condensed version) 205, (descriptions of scenery) 123-24, (essays) 140, (famine) 208, (figures) 124-26, (German translation) 143, (inns and hotels) 116-17, 122-23, (new edition) 143-44, (published) vii, (reviews) 141-43, 203; 'A Memory of Thomas Moore' 210;

Index

Retrospect of a Long Life 5, 17, 22, 25, 121-22, 201, 203; *The Use of Spiritualism* 22
Hamel the Obeah Man (Anon) 49
Hamilton, Elizabeth; *The Cottagers of Glenburnie* 96-98, 100; *Letters of a Hindoo Rajah* 96
Haydon, B.R., 'Death of the First Born' 28
Hayley, Barbara 32, 33, 150-51, 159
Hazlitt, William 75
Hemans, Felicia 14
Hickey, Anna Maria (daughter of Reverend William) 87
Hickey, Emily M. (granddaughter of Reverend William) 139
Hickey, Reverend William (pseud. Martin Doyle) 139, 87 148-50; *An Address to the Landlords of Ireland, etc.* 148; *Commonsense for Common People* 87; *Hints for the Small Farmers of the County of Wexford* 87; *Hints to the Small Holders and Peasantry of Ireland* 148; *Irish Cottagers* 97-98, 87
Hodgson and Graves (print publishers) 19
Hogg, James ('Ettrick Shepherd') 62
Holinshed, Raphael 113
Holland, Henry, *Travels in Albania* 2
Home, Daniel D., *Incidents in My Life* 22
Hood, Fanny (daughter of Thomas) 201, 202
Hood, Thomas 25, 201, 202
Hook, Theodore 75, 88
How and Parsons (publishers) 113
Howitt, Mary 25
Howitt, William 25

Illustrated News of the World 200
Inchbald, Mrs Elizabeth, *A Simple Story* 38
Ingoldsby, Thomas 200
Irish Builder 210, 211
Irish Farmer's and Gardener's Magazine and Register of Rural Affairs 87, 146, 150, 153, 155
Irish Farmers' Journal and Weekly Intelligencer 38, 61
Irish Life in Irish Fiction 211
Irish Monthly Magazine of Politics and Literature 33

Irish National Magazine 32
Irish Penny Magazine 34
Irish Short Story, The 212
Irish Times, The 210

James, G.P.R. 3
James II, King of England 130
Jeffares, Professor A.Norman 35
John Bull 9, 11, 18, 85, 86, 174
Johnson, Samuel, *Rasselas* 37
Juvenile Forget-me-Not (Annual) 26

Keating, Geoffrey 113
Keepsake (Annual) 27
Kennedy, Captain Pitt 127
Kilkenny Archaeological Society 211
Kilroy, James F.(ed.), *The Irish Short Story*, 212
King, Alice 200
Kingsley, Charles 44; *Yeast* 47
Knight, Charles 200
Koran 35
Kotzebue, August von *Travels in Italy* 1
Krans, Horatio 211

Lacey, Henry, *Life of David, King of Israel* 38
Lady's Magazine 64
Lamb, Charles 28, 200
Landon, Letitia E. 14, 28, 75, 200
Landseer, Sir Edwin Henry 29
le Fanu, Joseph 12
Leadbeater, Mary 98; *Cottage Dialogues among the Irish Peasantry* 97, 147-48; *Cottage Dialogues Among the Peasantry* 109
Lebow, Ned, 'British Images of Poverty in Pre-Famine Ireland' 55
Lever, Charles 11, 13, 34, 46, 188, 191-94; *Charles O'Malley* 191; *Harry Lorrequer* 177, 191; *Jack Hinton* 177, 191; *Lord Kilgobbin* 186; *The O'Donoghue* 177, 178-81, 184-86, 192, 194, 197; *St Patrick's Eve* 177, 182-84, 186; *Tom Burke of 'Ours'* 177
Lewes, G.H., *Rose, Blanche and Violet* 44
Lewis, Meriwether, and Clark, William, *Travels to the Source of the Missouri* 2
Lewis, George Cornewall 165; *On Local disturbances in Ireland* 163
Lewis, M.G. 3

Limerick, Siege of 130
Lind, Jenny (Mrs Otto Goldschmidt) 199, 203
Lister, T.H. 75
Literary Annuals *see* Annuals
Literary Gazette 8, 76, 98, 114; reviews *Ireland: its Scenery, etc.* 141; reviews *St Patrick's Eve* 183; reviews *Sketches of Irish Character* 20, 57, 62; reviews *The Whiteboy* 174
Literary Observer 18
Loan Fund System 114
Lorraine, Claude (painter) 2, 3
Lover, Mrs (friend of AMH) 23
Lover, Samuel 33; *Legends and Stories of Ireland* 69
Lyttleton, George 36

Mac Cuilleanain, Cormac 113
Macauley, Lord (Thomas Babington) 28
MacCarthy, B.G. 37
McGregor, Mr (local historian) 113
Mackintosh, Sir James 42
Maclise, Daniel 113, 213
MacManus, H. 113
McSkimmin, Mr, 'History of Carrickfergus' 113
Maginn, William 8, 28, 75
Maitland, Mrs, *The Ogilvies* 47; *Olive* 47
Malpas (Anon) 49
Mangan, James Clarence 34
Manwaring, Elizabeth, *Italian Landscape in Eighteenth Century England, etc.* 2-3
Martin, John 29
Martineau, Harriet 41, 42; *Deerbrook* 43-44, 45; *Illustrations of Political Economy* 43
Mary (Irish cook) 24
Mathew, Father Theobald (Capuchin friar) 58, 86, 128-29, 138, 173
Maturin, Charles 12; *The Wild Irish Boy* 168
Maynooth College 68, 85, 125, 143, 189
Memmi, Albert, *The Colonizer and the Colonized* 55
Mill, John Stuart 31
Mining Company of Ireland 114
Mitford, Mary Russell 25, 28, 213; 'A November Walk' 20; *Our Village* 61-64, 88; *Recollections* 63
Monthly Chronicle 10
Monthly Magazine 13, 39
Monthly Review 43
Moore, Thomas 7, 12, 113, 199-200; *Memoirs of Captain Rock* 49
More, Hannah 96; *Coelebs in Search of a Wife* 37
Morgan, Lady (Sydney Owenson) 3, 12, 43, 69, 145; and AMH 13-14, 110; *France* 1, 2; *The O'Briens and the O'Flahertys* 43, 49; *O'Donnell* 75; *The Wild Irish Girl* 168
Morning Journal 18
Moryson, Fynes 113
Mulready, William 29
Murray, H., *The Morality of Fiction* 38
Musgrave, Sir Richard 113-14

Nangle, Reverend Edward 143
Nation 142, 184, 186
National Gallery (London) 201
National Gallery of Ireland 200
New Monthly Magazine 75, 76; AMH writes for 31, 89; attitude to moral teaching 40, 41, 43, 49; reviews AMH 7, 10, 14, 62; SCH and 7, 18, 88
Newcomer, James, *Études Irlandaises* 211
Nicholl, A. 113
Nightingale, Florence 203
Nightingale Fund 203
North, Christopher 63
North British Review (Evangelical) 47
Norton, Caroline 75
novel 48; didactic 36-40, 44-45, 48-49, 197-98; English 45; French 45, 75; Gothic 69, 75; military 177, 186; philosophical 46; religious 44; of sensibility 37; 'silver fork' 75, 76, 88

Observer, The 9, 11, 202
O'Connell, Daniel 86, 173, 176, 192
O'Daly, Father, *History of the Geraldine Family* 113-14
Opie, Amelia 14, 41
Orange Order 186-87, 188, 191, 194-95, 196
Osborne, Richard Boyse 6; *Diary* 18, 110
Ostade, Adriaan van 51
Otway, Caesar 34, 69

Index

Paine, Thomas 96
Park, Mungo 1
Penal Laws 129, 130
Pepys, Samuel, diary 76
Petrie, Charles 34
Petrie, George 114
Pilot 173, 176
Poor Law 63, 89-90, 144, 208
Poor Law Commissioners 114
Poussin, Nicholas 2, 3
Power, Tyrone 9, 80
Precaution (Anon) 43
Prophetess, The (Anon) 49
Prospective Review 47
Prout, Samuel 29, 113
Public Health Act (1841) 27
Punch 51, 202
Purnell, Mr 139

Quarterly Review 41-42, 43

Radcliffe, Mrs Ann 3
Rainsford, Captain Marcus, *Account of Haiti* 1
Recluse, The (Anon) 41
Reeves, Sim 200
Religious Tract Society 96
Report of the Commissioners on Municipal Corporations in Ireland 114
Report of the Select Parliamentary Committee (1823) on the Condition of the Poor in Ireland 114, 133
Ribbonmen 131, 189, 196
Richardson, Samuel; *Clarissa* 37; *Pamela* 37
Robertson, John 31
Roche, Alexander D. 77
Rosa, Salvator 2-3
Ross, Angus 37
Rousseau, Jean-Jacques, *La Nouvelle Héloïse* 40
Royal Dublin Society 33, 87
Royal Irish Academy 33
Ruskin, John 29
Ryland, Reverend Mr (local historian) 113
Ryle, Gilbert 38

St James's Magazine 200, 202
Savage, M.W., *Reuben Medlicott* 45
Schiller, Friedrich von 48

Schirmer, Gregory, 'Tales from Big House and Cabin' 212
Schoberl, Frederick (editor) 27, 76
Scott, Michael 205, 213
Scott, Sir Walter 3, 28, 40, 46
Senior, Mrs Henry 22
Shanavests 131
Sharpe's Magazine 202
Sloan, Barry; 'Mrs Hall's Ireland' 212; *Pioneers of Anglo-Irish Fiction 1800-1850* 212
Smith, Dr (local historian) 113
Society of Antiquaries 120
Society for the Diffusion of Useful Knowledge 96
Southey, Robert 28
Spectator, The 9, 80-81, 173, 182, 183
Spirit and Manners of the Age 18, 19-20, 25
Steen, Jan 51
Stendhal (pseud. of Henri Marie Beyle), *Rome, Naples and Florence* 2
Stephenson, Dr, *Brief History of Grey Abbey in County Down* 114
Style, Sir Charles 127
Sunday Times, The 9, 10, 11, 141, 173-74, 183

Tablet 173, 192
Tahitian and Irish Spelling Books 4
Tait's Edinburgh Magazine 40, 188, 190, 213
Talmud 35
Taylor, Jane, *Display* 41
temperance movement 128, 210
Thackeray, William Makepeace 12, 28
Thomson, James, *Seasons* 36
Three Hundred Years of Irish Periodicals 32
Times, The 62, 89, 210
Town 18
Trollope, Anthony 12; *The Macdermots of Ballycloran* 206
Turnbull, John, *Voyage round the World* 1
Turner, Joseph Mallord William 29
Tusser, Thomas 36
Tutor's Ward, The (Anon) 45
Twomey, Mr J.C. 211

United Irishmen 194
Vancouver, Captain George 1
Vernon, Robert (art collector) 201

Victoria, Queen of Great Britain and Ireland 201
Virgil, *The Georgics* 44
Virtue, George (publisher) 143
Vizetelly, Henry (publisher) 139; *The Later Years* 199

Walsh, Reverend Robert 113
Ward, Robert Plumer 75, 200; *Tremaine* 41
Watchman (Methodist newspaper) 18
Webster, Grace 14
Webster, James, *Travels through India and the Crimea* 2
Weekly Dispatch 9-10, 58
Weigall, C.H. 113
Wesley, John 1, 37
Westley and Davis (London publisher) vii, 19, 20, 27, 75
Westminster Review 31, 44, 49

Wexford Independent 210-11
Whiteboys 131, 163, 164, 165-66, 172, 194-95
Wilberforce, William 19
Wilkie, Sir David 29
William III, King of England, Prince of Orange 130
Wilson, John, *Lights and Shadows of Scottish Life* 77
Windele, Mr (local historian) 113
Wolff, Robert Lee 212
Wood, Mrs Henry 200
Woodroffe, Dr (local historian) 113
Wordsworth, William 28, 200

Young, Arthur 1
Young Ireland Party 173
Young Ireland Rebellion 207

Zschokke, Heinrich 40